Math[illegible]
for Li[illegible]

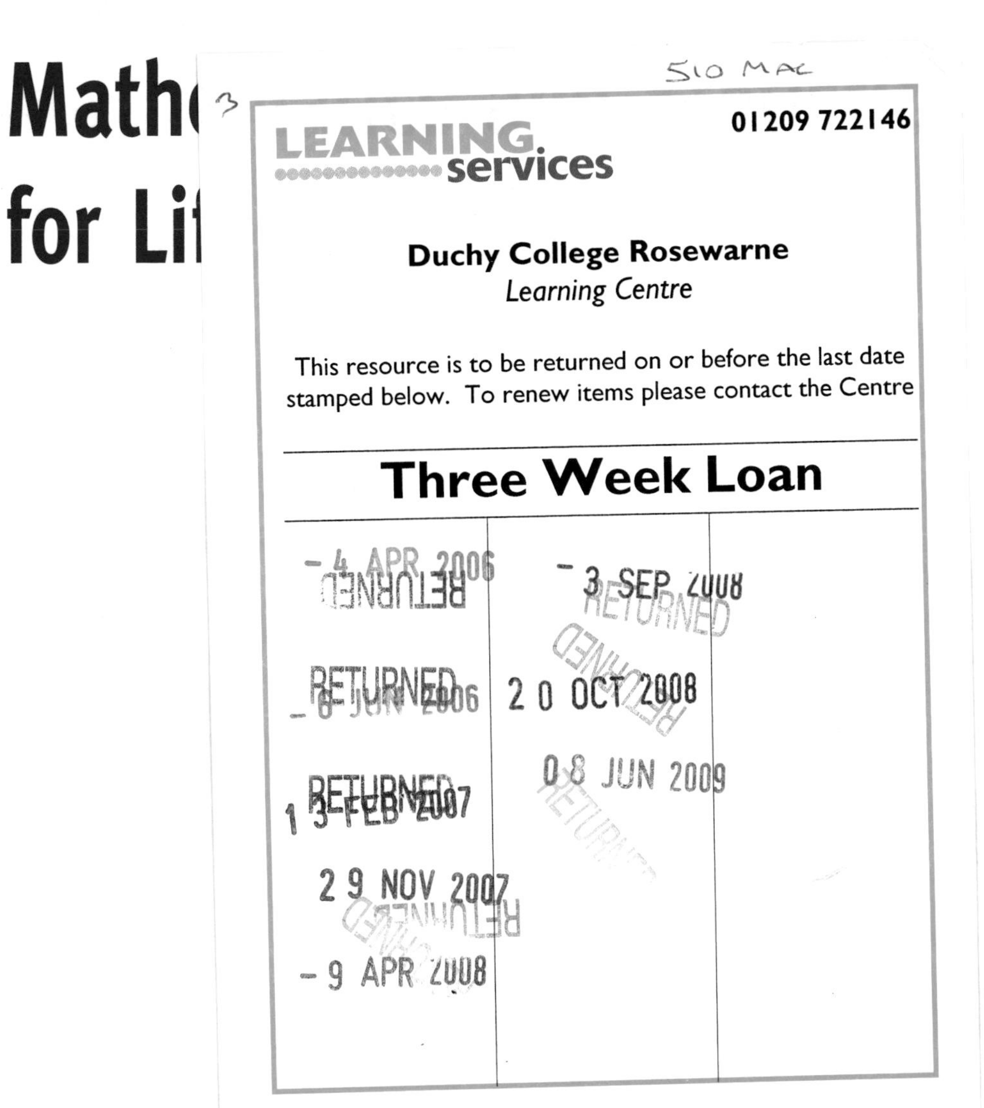

BIOS INSTANT NOTES

Series Editor: B.D. Hames, School of Biochemistry and Molecular Biology, University of Leeds, Leeds, UK

Biology

Animal Biology, Second Edition
Biochemistry, Third Edition
Bioinformatics
Chemistry for Biologists, Second Edition
Developmental Biology
Ecology, Second Edition
Genetics, Second Edition
Immunology, Second Edition
Mathematics & Statistics for Life Scientists
Medical Microbiology
Microbiology, Second Edition
Molecular Biology, Third Edition
Neuroscience, Second Edition
Plant Biology, Second Edition
Sport & Exercise Biomechanics
Sport & Exercise Physiology

Chemistry

Consulting Editor: Howard Stanbury

Analytical Chemistry
Inorganic Chemistry, Second Edition
Medicinal Chemistry
Organic Chemistry, Second Edition
Physical Chemistry

Psychology

Sub-series Editor: Hugh Wagner, Dept of Psychology, University of Central Lancashire, Preston, UK

Cognitive Psychology
Physiological Psychology
Psychology
Sport & Exercise Psychology

Mathematics and Statistics for Life Scientists

Dr Aulay Mackenzie

Department of Biological Sciences,
University of Essex, Colchester, UK

Published by:
Taylor & Francis Group
In US: 270 Madison Avenue
New York, NY 10016
In UK: 4 Park Square, Milton Park
Abingdon, OX14 4RN

First published 2005

ISBN: 1-8599-6292-0

A catalog record for this book is available from the British Library.

Library of Congress Cataloging-in-Publication Data

Mackenzie, A. (Aulay)
Mathematics and statistics for life scientists / Aulay Mackenzie.
p. cm. -- (BIOS instant notes)
Includes index.
ISBN 1-85996-292-0
1. Mathematics -- Textbooks. 2. Life sciences -- Mathematics -- Textbooks.
3. Life sciences -- Statistical methods -- Textbooks. I. Title. II. Series.
QA37.3.M33 2005
510--dc22 2005022851

Editor: Elizabeth Owen
Editorial Assistant: Chris Dixon
Production Editor: Georgina Lucas/Simon Hill
Typeset by: Phoenix Photosetting, Chatham, Kent, UK
Printed by: TJ International Ltd, Padstow, Cornwall

Printed on acid-free paper

10 9 8 7 6 5 4 3 2 1

Taylor & Francis Group
is the Academic Division of T&F Informa plc.

Visit our web site at http://www.garlandscience.com

CONTENTS

Section A – Using numbers in life sciences		**1**
Section B – Measures and units		**3**
B1	Types of measurement	3
B2	Units: The Système International	6
B3	Preparing solutions	10
Section C – Handling and presenting data		**15**
C1	Handling data	15
C2	Presenting data	20
Section D – Building blocks of mathematics		**25**
D1	Manipulating numbers: algebra	25
D2	Trigonometry	29
D3	Indices and logarithms	33
Section E – Using mathematics		**39**
E1	pH, Beer's law, and scaling	39
E2	Defining biological relationships	46
Section F – Rates of change: Differentiation		**55**
F1	Finding gradients and rates	55
F2	Other functions	59
Section G – Rates of change Integration		**65**
G1	Integration and integrals	65
G2	Position, velocity, and acceleration	71
G3	Methods of integration	76
G4	Areas under lines	87
G5	Numerical integration	90
Section H – Equations		**95**
H1	Differential equations	95
H2	Difference equations	106
Section I – Using equations		**113**
I1	Population growth	113
I2	Heat loss from a body	125
I3	Chemical kinetics	127
Section J – Building blocks of statistics		**135**
J1	Why use statistics?	135
J2	Experimental design	138
J3	Tests and testing	143

Section K – Finding the right statistical test **149**
K1 Searching for patterns and causes 149
K2 Searching for patterns in continuous data 157
K3 Searching for patterns in count data 165
K4 Searching in a data pond 168

Index **173**

A1 USING NUMBERS IN THE LIFE SCIENCES

Key Notes

Quantification	The interpretation of data, diagnosis, analysis and deductions in scientific study cannot be made without measurements of some type.
Mathematics	The relationships between quantities, magnitudes and forms are dealt with in mathematics.
Statistics	The detection of patterns in data sets is facilitated by statistics.
Biological data	The vast diversity in biological systems causes the data to be 'noisy'. Therefore detecting patters requires rigorous analysis.

Quantification

Mathematics and statistics are important in the life sciences because scientific study of any type depends on **quantification**. It is rarely possible to be able to make any interpretation, diagnosis, analysis or deduction without measurements or counts or some kind. So, whilst your main interest may be immunology or coral reef ecology, nutrition or biotechnology, you need to have a grasp of the basic tools to handle numbers. This book aims to give you a basic toolkit in an accessible and digestible format to help your forays into your chosen field.

Mathematics

The group of related sciences which includes algebra, geometry and calculus is known as **mathematics**. The inter-relationships between numbers, quantities, shapes and space are studied by using special notation.

Statistics

The detection of patterns in large data sets, such as frequently encountered in the life sciences, is facilitated by **statistics**, which is a branch of mathematics rather than a separate discipline. Crudely, the role of general mathematics can be described as 'taming' the numerical patterns in data, while statistics interprets the patterns.

Biological data

Biological systems differ from some other sciences, such as engineering and physics, in that the genetic variation between individual organisms, and between populations and species creates a vast diversity in responses and patterns. This means that **biological data** has a tendency to be 'noisy'. This is particularly true in ecological studies, where multiple biological and other environmental factors may be impacting upon the data set. For this reason, rigorously detecting patterns in biological data requires particularly careful analysis, and indeed the field of statistics has partly grown out of the analysis required for biological applications.

The topics in this book cover the basic needs of a life science undergraduate, but the book is not intended to be comprehensive. I have tried to avoid coverage of topics or levels of detail that are not wholly necessary: nothing is here on the basis that it will 'do you good', but there will be some areas and some students which require more advanced coverage than this.

I have adopted two somewhat different approaches to the mathematical and statistical components. For the mathematics, it is important to understand what the procedures do and how they work in order to be able to implement them, so there is step-by-step coverage. In contrast, for the statistics, it is important to know what a particular test does, but not to deal with the mathematical underpinnings. A simple '**black box**' approach is taken – that is to say, the inner workings of the mathematical mechanics of the statistical tests are not covered, except where absolutely necessary. The mission is to help you locate the right test, apply it correctly and understand the result. Just as effectively driving a car does not require intimate mechanical knowledge of the engine, so statistical tests can be properly and effectively used without a detailed understanding of what makes them tick. My underlying assumption is that you will use statistical software to perform the calculations, and I do not adhere to the viewpoint that undertaking these by hand would make you better at understanding the test, any more than taking apart a car engine will make you a better driver. If you go on to develop a deeper understanding of statistics, you may find a more mechanistic understanding is useful. Because I assume you will be using software for analysis, there is no appendix of probability tables.

The statistical tests covered are those which are in commonest use and have the broadest application. It may be that you will come across situations that require tests beyond these, but this should not be a common occurrence. In any case, once you have a grasp of the tests described here, you will be well equipped to tackle others. Also, a brief coverage of multivariate methods is included, as much to provide a guide to interpreting these analyses when presented elsewhere as to tackling them yourself.

B1 TYPES OF MEASUREMENT

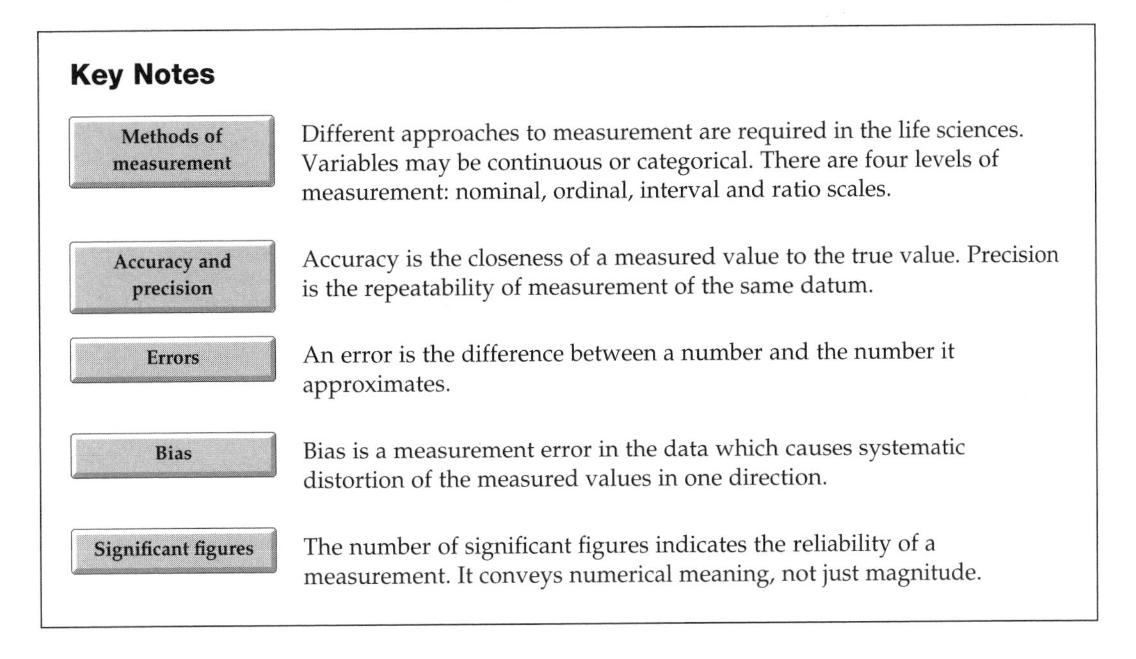

Key Notes

Methods of measurement	Different approaches to measurement are required in the life sciences. Variables may be continuous or categorical. There are four levels of measurement: nominal, ordinal, interval and ratio scales.
Accuracy and precision	Accuracy is the closeness of a measured value to the true value. Precision is the repeatability of measurement of the same datum.
Errors	An error is the difference between a number and the number it approximates.
Bias	Bias is a measurement error in the data which causes systematic distortion of the measured values in one direction.
Significant figures	The number of significant figures indicates the reliability of a measurement. It conveys numerical meaning, not just magnitude.

Methods of measurement

Applying measurements appropriately in all sciences is a key skill. In the life sciences data comes in a range of forms that require different approaches to measurement. Some variables are **continuous** (or **metric**), which fall along an uninterrupted scale (e.g. length, mass, temperature) whilst count data can only be whole numbers and are **categorical** (also termed **discrete, discontinuous** or **nonmetric**). Four levels of measurement are recognized: the **nominal, ordinal, interval** and **ratio** scales.

A **nominal scale** classifies observations into exclusive categories that have no relative rank. Examples include species, gender, color and habitat type.

An **ordinal scale** classifies observations into ranked exclusive categories. Examples include abundance scales (e.g. the DAFOR scale (**d**ominant, **a**bundant, **f**requent, **o**ccasional, **r**are)) used to record the abundance of plant species in a quadrat) and developmental status (e.g. newborn, juvenile, yearling, adult).

An **interval scale** defines observations on a continuous scale that has no absolute zero. Examples include the Celsius temperature scale and the calendar date. (Note that the absence of an absolute zero means it is not true to say that 15°C is three times warmer than 5°C, nor that birds arriving on the 8th May are twice as late as those arriving on the 4th May.)

A **ratio scale** defines observations on a continuous scale that has an absolute zero. Examples include length, weight and most physical measurements. Thus, in contrast to measurements made in the interval scale, it is true to say that a plant leaf of length 45 mm is half the length of one of 90 mm length.

Accuracy and precision

When measurements are being made, it is necessary that they be accurate or precise or both. Unfortunately, the terms 'accuracy' and 'precision' are often colloquially used interchangeably, but the distinction in scientific use is important. **Accuracy** is the closeness of a measured value to the true value. **Precision** is the repeatability of measurements of the same datum (i.e. the closeness of measured values to one another) (see *Fig. 1*). It is possible for a **bias** in the measurement apparatus to give rise to very precise, repeatable values which were inaccurate.

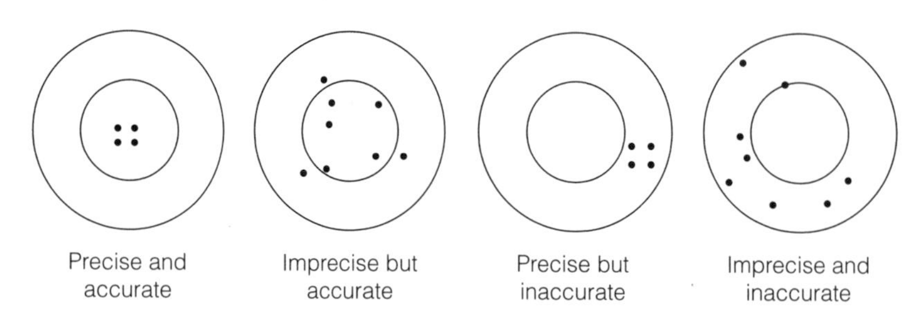

Fig. 1. Target diagrams of precision and accuracy.

Errors

Error is the difference between a number and the number it approximates. It is impossible to remove all error from estimates, but careful scientific procedure can limit it. If we wish to estimate the mean length of a species of bumble bee, we may choose to measure thirty specimens and take the mean value of this sample. Two sources of error would arise: **measurement error**, due to problems with handling and using the measurement equipment, and **sampling error**. Sampling error arises because only a subset of the whole population is being measured. There is a third source of error, **rounding error**, that arises as a result of numbers that have been calculated with imprecise subcomponents or parameters. An example of this is given in the example cited in Topic D3 (see p. 36).

To achieve accurate results, it is important to minimize all sources of error. Measurement error is minimized by using carefully standardized procedures and accurate instruments. Sampling error declines as the sample size increases, so larger samples are always better, but are more costly in terms of time and effort. Sampling procedures and choosing the best compromise in sample size are discussed in Topic K2.

Bias

Bias is a type of **measurement error** in the data which causes **systematic** distortion of the measured values in one direction. Bias can be caused by defective or incorrectly calibrated instruments, by the effects of experimental manipulation or preparation, or by the observer anticipating particular results and interpreting data with this perspective (e.g. by ignoring data points which disagree with this preconception). To minimize the occurrence of bias, carefully standardized procedures should be adopted with well calibrated instruments. In experiments where observer bias may be strong (e.g. in drug trials involving placebos), 'blind' (subjects who are unaware of their treatment regime) or 'double-blind' (both the subjects and observers are unaware of the treatment regime) approaches may be useful. Bias can be detected by measuring a variable in the same set of samples in different ways and checking for consistent results.

Significant figures

It is common for numbers to be calculated which, in their raw state, erroneously appear to indicate very high levels of accuracy. Stating these numbers to a limited number of significant figures reduces the implied accuracy levels to defendable levels. The number of significant figures indicates the reliability of the measurement and is the number of digits in the number which convey numerical meaning, not just magnitude. So, the statement '100 kg to three significant figures' means that the accuracy is stated as being to within 1 kg of 100 kg, whilst '100 kg to two significant figures' means that the true value may lie between 95 kg and 105 kg. Similarly, '0.010 µm to two significant figures' means that the value lies between 0.009 5 µm and 0.010 5 µm.

Example
A compact disc on my desk has a radius of 60 mm, to the nearest millimeter. As the circumference of a circle is given by πr, where the value of the constant, π, is 3.141 592 65 …, the circumference of the disc is therefore 3.141 592 65 … × 60 mm, which gives 188.495 559 2 … mm. As the initial measurement was only accurate to the nearest millimeter, it is clearly nonsensical to state a derived value as having far higher accuracy. As the measurement was taken to two significant figures, so the circumference of the disc should be stated as 190 mm (to three significant figures).

General suggested rules

- Calculations involving measured quantities should be **no more precise** than the **least precise measurement**.
- Means should be stated to a level of accuracy one significant figure greater than that used in the raw data.
- **Standard deviations** should be stated to a level of accuracy two significant figures greater than that used in the raw data.

B2 UNITS: THE SYSTÈME INTERNATIONAL

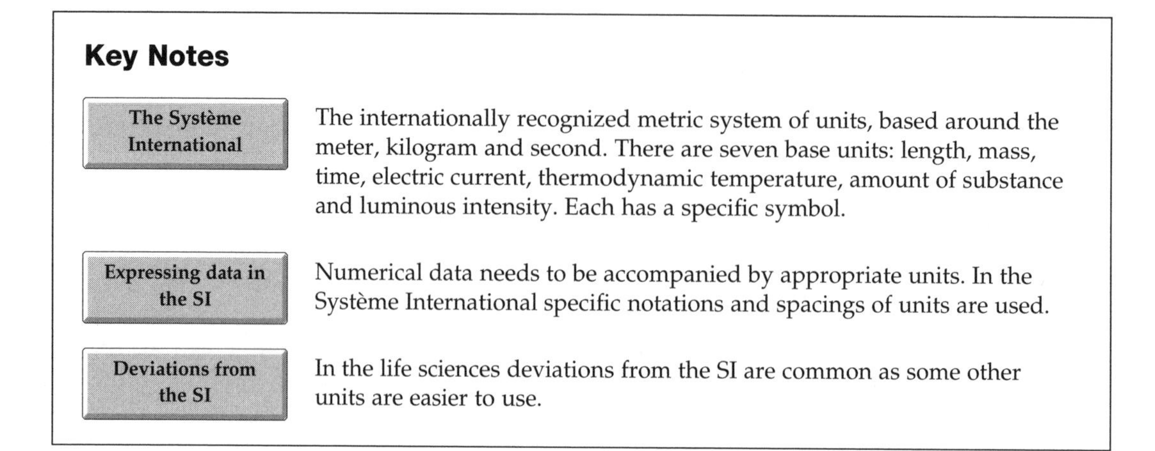

Key Notes

The Système International — The internationally recognized metric system of units, based around the meter, kilogram and second. There are seven base units: length, mass, time, electric current, thermodynamic temperature, amount of substance and luminous intensity. Each has a specific symbol.

Expressing data in the SI — Numerical data needs to be accompanied by appropriate units. In the Système International specific notations and spacings of units are used.

Deviations from the SI — In the life sciences deviations from the SI are common as some other units are easier to use.

The Système International

Having a single unified system of units greatly aids the communication of data and scientific understanding and also simplifies calculations. The **Système International d'Unités**, usually shortened to **SI**, is the internationally recognized metric system, based around the meter, kilogram and second.

The SI system consists of **seven base units** and two supplementary units (*Table 1*); each of these has a specific symbolic abbreviation. These **fundamental units** can be combined to give **compound units**, of which the more commonly used are given special symbols (*Table 2*).

Prefixes denote multiplication factors of 10^3 and are used to make large or small numbers manageable (*Table 3*).

Table 1. The base and supplementary SI units

Measured quantity	Name of SI unit	Symbol	
Base units			
Length	meter	m	
Mass	kilogram	kg	
Amount of substance	mole	mol	
Time	second	s	
Electric current	ampere	A	
Temperature	kelvin	K	(note: NOT °K)
Luminous intensity	candela	cd	
Supplementary units			
Plane angle	radian	rad	
Solid angle	steradian	sr	

Table 2. Some important derived SI units

Measured quantity	Name of unit	Symbol	Definition in base units	Alternative in derived units
Force	newton	N	m kg s^{-2}	J m^{-1}
Energy	joule	J	m^2 kg s^{-2}	N m
Pressure	pascal	Pa	kg m^{-1} s^{-2}	N m^{-2}
Power	watt	W	m^2 kg s^{-3}	J s^{-1}
Radioactivity	becquerel	Bq	s^{-1}	
Luminous flux	lumen	lm	cd sr	
Illumination	lux	lx	cd sr m^{-2}	lm m^{-2}
Frequency	hertz	Hz	s^{-1}	
Enzyme activity	katal	kat	mol substrate s^{-1}	
Electric potential difference	volt	V	m^2 kg A^{-1} s^{-3}	W A^{-1}
Electric resistance	ohm	Ω	m^2 kg A^{-2} s^{-3}	V A^{-1}
Electric conductance	siemens	S	s^3 A^2 kg^{-1} m^{-2}	A V^{-1} or Ω^{-1}

Table 3. Prefixes used in the SI

Multiple	Prefix	Symbol
10^{-3}	milli	m
10^{-6}	micro	μ
10^{-9}	nano	n
10^{-12}	pico	p
10^{-15}	femto	f
10^{-18}	atto	a
10^{3}	kilo	k
10^{6}	mega	M
10^{9}	giga	G
10^{12}	tera	T
10^{15}	peta	P
10^{18}	exa	E

Expressing data in the SI

When data are expressed using the SI it always needs to be accompanied by the appropriate units (except in unusual cases such as pH and dimensionless ratios).

The SI incorporates a particular spacing and use of symbols. For example, a measurement of 302 cm (to the nearest centimeter) should be expressed as 3.02 m. Note that

- there is a space between the number and the units,
- there is no full stop to indicate abbreviation,
- units are always in singular form.

Thus 3.02m, 3.02 m. and 3.02 ms are all incorrect.

In the case of **compound units**, for example if the velocity of a moving animal was recorded as 1.56 meters per second, this is correctly expressed as 1.56 m s^{-1}. Note that

- there is a space between the separate symbols,
- that a negative power is used, not a solidus (/).

Thus 1.56 ms^{-1} and 1.56 m/s are both incorrect.

In the SI, **prefixes** should be used to denote multiples of 10^3 so the numbers are always kept between 0.1 and 1 000. Thus a measurement of 6 241 m should be expressed as 6.241 km. Note that there is no space between the prefix and the unit symbol.

Deviations from the SI

In the life sciences deviations from the SI are common: previously used, or more convenient, units for volume, concentration, temperature, time and light are encountered frequently.

Strictly the SI unit of **volume** is the cubic meter, m^3. However, use of the liter (l or L) and milliliter (ml or mL) is still widespread in many laboratory applications. In calculations it may be most convenient to convert liters to SI units: 1 liter $= 1 \times 10^{-3}\,m^3$.

Details concerning **concentration** are dealt with extensively in Topic B3.

The SI unit of **temperature** is the **kelvin** (K) but this is rarely used in everyday life science applications where the **Celsius** scale (°C) is employed. This scale has interval steps of the same magnitude (i.e. a 1 K increase in temperature is exactly the same as 1°C increase), but the zero point is at 275.15 K, the melting point of pure ice at standard pressure. (Standard temperature and pressure (STP) are 293.15 K and 0.101 325 MPa, respectively.)

In SI the base unit for **time** is the **second** (s), and this should be used where possible. Nevertheless, longer time scales may be better such as hours (h), days (d) and years (yr) may be more appropriate in some situations.

The SI base unit for luminous intensity (**light**), is the candela (cd) and the derived units the lumen and lux. These are rather substandard compared to the high precision possessed by the other base units. These units are derived from the response to light of the human eye (from a small sample of 52 US soldiers in 1923). As light sources differ in their spectral qualities and biological receptors (e.g. eyes, skin, plant chloroplasts) differ in their spectral sensitivities, the candela has rather limited value in studies not primarily concerned with human vision. Therefore it is preferable to express light either in terms of energy content (e.g. $W\,m^{-2}$) or photon density (e.g. $\mu mol\,m^{-2}\,s^{-1}$), and it is often valuable to specify the spectrum involved.

It may be necessary to convert between non-SI and SI units, for example to allow comparison with results in the older literature. *Table 4* gives a range of conversion factors.

To convert from non-SI to SI, multiply the number in the old units by the conversion factor.

Example

To convert square inches to square meters:

$$1\,452\ in^2 = 1\,452 \times (645.16 \times 10^{-6}) = 0.936\,77\ m^2$$

To convert a number in SI units to the non-SI form, multiply by the reciprocal.

Example

To convert kilograms to ounces:

$$2.314\ kg = 2.314 \times 35.274\,0 = 81.624\,0\ oz.$$

Table 4. SI Conversion factors

Parameter		SI unit	Abbreviation	Conversion	Reciprocal
Area					
acre	⇒	square meter	m^2	$4.046\,86 \times 10^3$	$0.247\,105 \times 10^{-3}$
hectare (ha)	⇒	square meter	m^2	10×10^3	0.1×10^{-3}
square foot (ft^2)	⇒	square meter	m^2	0.092 903	10.763 9
square inch (in^2)	⇒	square meter	m^2	645.16×10^{-6}	$1.550\,00 \times 10^3$
square yard (yd^2)	⇒	square meter	m^2	0.836 127	1.195 99
Energy					
erg	⇒	joule	J	0.1×10^{-6}	10×10^6
kilowatt hour (kWh)	⇒	joule	J	3.6×10^6	$0.277\,778 \times 10^{-6}$
Length					
Ångstrom (Å)	⇒	meter	m	0.1×10^{-9}	10×10^9
Foot (ft)	⇒	meter	m	0.304 8	3.280 84
Inch (in)	⇒	meter	m	25.4×10^{-3}	39.370 1
Mile	⇒	meter	m	$1.609\,34 \times 10^3$	$0.621\,373 \times 10^{-3}$
Yard (yd)	⇒	meter	m	0.914 4	1.093 61
Mass					
Ounce (oz)	⇒	kilogram	kg	$28.349\,5 \times 10^{-3}$	35.274 0
Pound (lb)	⇒	kilogram	kg	0.4535 92	2.204 62
Stone	⇒	kilogram	kg	6.350 29	0.157 473
Hundredweight (cwt)	⇒	kilogram	kg	50.802 4	$19.684\,1 \times 10^{-3}$
UK ton	⇒	kilogram	kg	$1.016\,05 \times 10^3$	$0.984\,203 \times 10^{-3}$
Pressure					
Atmosphere (atm)	⇒	pascal	Pa	101 325	$9.869\,23 \times 10^{-6}$
Bar (b)	⇒	pascal	Pa	100 000	10×10^{-6}
millimeter of mercury/mmHg	⇒	pascal	Pa	133.322	$7.500\,64 \times 10^{-3}$
torr (Torr)	⇒	pascal	Pa	133.322	$7.500\,64 \times 10^{-3}$
Radioactivity					
Curie (Ci)	⇒	becquerel	Bq	37×10^9	$27.027\,0 \times 10^{-12}$
Temperature					
centigrade (Celsius) degree/°C	⇒	kelvin	K	K – 273.15	°C + 273.15
Fahrenheit degree (°F)	⇒	kelvin	K	(K × 9/5) – 459.67	(°F + 459.67) × 5/9
Volume					
cubic foot (ft^3)	⇒	cubic meter	m^3	0.028 316 8	35.314 7
cubic inch (in^3)	⇒	cubic meter	m^3	$16.387\,1 \times 10^{-6}$	$61.023\,6 \times 10^3$
cubic yard (yd^3)	⇒	cubic meter	m^3	0.764 555	1.307 95
UK pint (pt)	⇒	cubic meter	m^3	$0.568\,261 \times 10^{-3}$	1 759.75
US pint (liq pt)	⇒	cubic meter	m^3	$0.473\,176 \times 10^{-3}$	2 113.38
UK gall (gal)	⇒	cubic meter	m^3	$4.546\,09 \times 10^{-3}$	219.969
US gall (gal)	⇒	cubic meter	m^3	$3.785\,41 \times 10^{-3}$	264.172

B3 PREPARING SOLUTIONS

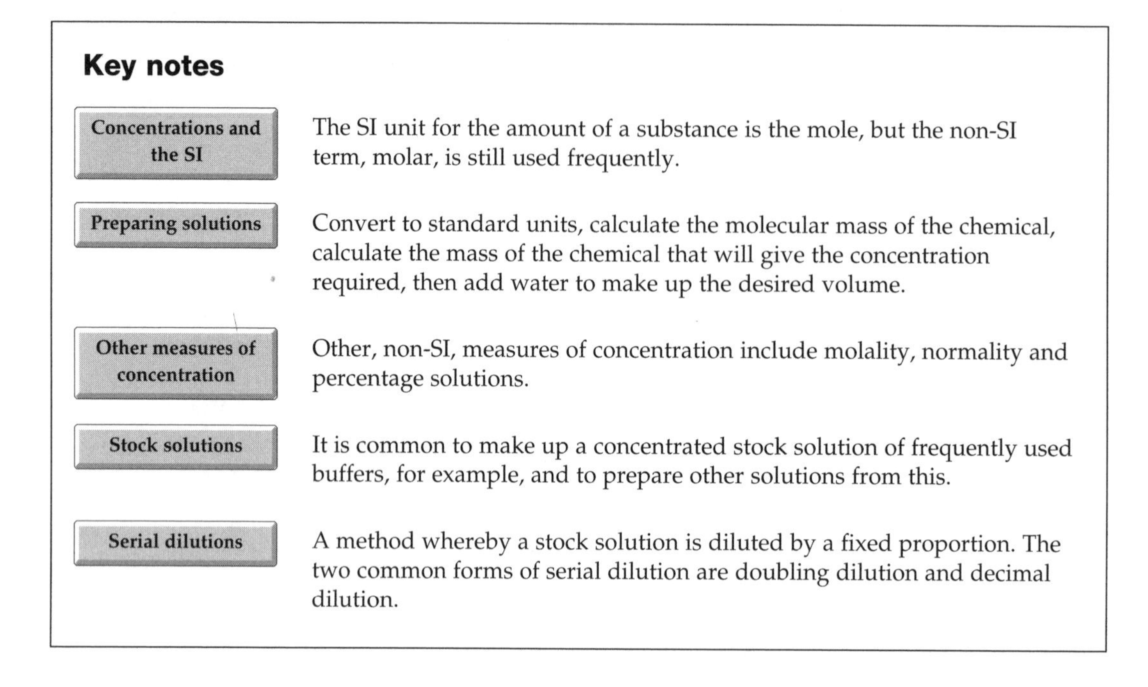

Key notes

Concentrations and the SI	The SI unit for the amount of a substance is the mole, but the non-SI term, molar, is still used frequently.
Preparing solutions	Convert to standard units, calculate the molecular mass of the chemical, calculate the mass of the chemical that will give the concentration required, then add water to make up the desired volume.
Other measures of concentration	Other, non-SI, measures of concentration include molality, normality and percentage solutions.
Stock solutions	It is common to make up a concentrated stock solution of frequently used buffers, for example, and to prepare other solutions from this.
Serial dilutions	A method whereby a stock solution is diluted by a fixed proportion. The two common forms of serial dilution are doubling dilution and decimal dilution.

Concentrations and the SI

It is commonly necessary in laboratory science to create chemical solutions of known concentrations. The SI unit for the amount of a substance is the mole (mol). The SI derived unit of concentration is thus mol m^{-3}. However, the use of the non-SI term, the molar (M) solution, equivalent to 1 mole in 1 liter of water, is widespread, as is 'millimolar' (mM). Given that in SI 'M' is a prefix (mega), there is some possibility of confusion here, so care must be taken. A solution with a concentration of 1 kmol m^{-3} is equivalent to a 1 M solution, and 1 mol m^{-3} is equivalent to a 1 mM solution.

Some **definitions** are useful. The **mole** is the SI unit of substance. 1 mole of a chemical compound has a mass equal to its **molecular weight** in grams. A mole of a chemical compound contains 6.022×10^{23} molecules (this number is known as Avogadro's number). Moles can also be used to express the amount of atoms and ions, for example.

Molecular weight is the sum of the atomic weights (shown on the periodic table (*Fig. 1*)) of all the atoms in a molecule.

Molarity is a measure of solution concentration, expressed in moles per liter (mol l^{-1}). Thus, the molecular weight of a compound dissolved in 1 liter of water gives a concentration of 1 mol l^{-1} (also called 'a 1 molar solution').

The **solute** is the substance dissolved in liquid in a solution. A **solvent** is a liquid capable of dissolving another substance. Unless stated otherwise, it is assumed in life science applications that the solvent is water.

Atomic number

6 C 12.01

Symbol

Atomic weight

	1	2	3	4	5	6	7	8	9	10	11	12	13	14	15	16	17	18
1	1 **H** 1.008																	2 **He** 4.003
2	3 **Li** 6.941	4 **Be** 9.012											5 **B** 10.81	6 **C** 12.01	7 **N** 14.01	8 **O** 16.00	9 **F** 19.00	10 **Ne** 20.18
3	11 **Na** 22.99	12 **Mg** 24.31											13 **Al** 26.98	14 **Si** 28.09	15 **P** 30.97	16 **S** 32.07	17 **Cl** 35.45	18 **Ar** 39.95
4	19 **K** 39.10	20 **Ca** 40.08	21 **Sc** 44.96	22 **Ti** 47.88	23 **V** 50.94	24 **Cr** 52.00	25 **Mn** 54.94	26 **Fe** 55.85	27 **Co** 58.93	28 **Ni** 58.69	29 **Cu** 63.55	30 **Zn** 65.39	31 **Ga** 69.72	32 **Ge** 72.61	33 **As** 74.92	34 **Se** 78.96	35 **Br** 79.90	36 **Kr** 83.80
5	37 **Rb** 85.47	38 **Sr** 87.62	39 **Y** 88.91	40 **Zr** 91.22	41 **Nb** 92.91	42 **Mo** 95.94	43 **Tc** 98.91	44 **Ru** 101.1	45 **Rh** 102.9	46 **Pd** 106.4	47 **Ag** 107.9	48 **Cd** 112.4	49 **In** 114.8	50 **Sn** 118.7	51 **Sb** 121.8	52 **Te** 127.6	53 **I** 126.9	54 **Xe** 131.3
6	55 **Cs** 132.9	56 **Ba** 137.3	71 **Lu** 175.0	72 **Hf** 178.5	73 **Ta** 180.9	74 **W** 183.8	75 **Re** 186.2	76 **Os** 190.2	77 **Ir** 192.2	78 **Pt** 195.1	79 **Au** 197.0	80 **Hg** 200.6	81 **Tl** 204.4	82 **Pb** 207.2	83 **Bi** 209.0	84 **Po** 209.0	85 **At** 210.0	86 **Rn** 222.0
7	87 **Fr** 223.0	88 **Ra** 226.0	103 **Lr** 262.1	104 **Rf** 261.1	105 **Db** 262.1	106 **Sg** 263.1	107 **Bh** 264.1	108 **Hs** 265.1	109 **Mt** 268	110 **Uun** 269	111 **Uuu** 272	112 **Uub** 277		114 **Uuq** 289		116 **Uuh** 292		118 **Uuo** 293

6	57 **La** 138.9	58 **Ce** 140.1	59 **Pr** 140.9	60 **Nd** 144.2	61 **Pm** 146.9	62 **Sm** 150.4	63 **Eu** 152.0	64 **Gd** 157.3	65 **Tb** 158.9	66 **Dy** 162.5	67 **Ho** 164.9	68 **Er** 167.3	69 **Tm** 168.9	70 **Yb** 173.0
7	89 **Ac** 227.0	90 **Th** 232.0	91 **Pa** 231.0	92 **U** 238.0	93 **Np** 237.0	94 **Pu** 244.1	95 **Am** 243.1	96 **Cm** 247.1	97 **Bk** 247.1	98 **Cf** 251.1	99 **Es** 252.0	100 **Fm** 257.1	101 **Md** 258.1	102 **No** 259.1

Fig. 1. Periodic table of the elements.

Preparing solutions

To make up a solution of given volume and concentration:

- convert to standard units if required: express the volume in liters and concentration in mol l^{-1},
- calculate the molecular mass (in grams) of the chemical,
- calculate the mass of chemical (in grams) that will give the concentration required:

$$\text{mass of chemical} = \text{volume} \times \text{molecular mass} \times \text{concentration},$$

- Follow a theoretical or practical approach:
  - For the theoretical approach add water to the chemical to make up to the desired volume.
  - For the practical approach add the measured mass of the chemical to a beaker (or other vessel), then about 80% of the final solution volume of water. Dissolve thoroughly (use a magnetic stirrer and heat if required). Adjust the pH if necessary, then make up to the correct volume in a volumetric flask or measuring cylinder.

In practical laboratory work it is often difficult to accurately weigh out very small masses, particularly if the chemical is composed of large crystals or is 'sticky'. In this case it may be appropriate to either (1) make up a larger volume; or (2) make up a more concentrated stock solution and dilute this (see **diluting** below); or (3) weigh the mass first then calculate the volume of water required to make the desired concentration (see Example 2 below).

Example 1

How much NaOH is required to make 50 ml of a 5 mM solution?

(1) Convert to the standard units: volume = 0.050 l, concentration = 5×10^{-3} mol l^{-1}.

(2) The molecular mass (in grams) of NaOH is $22.99 + 16.00 + 1.008 = 39.998$ g.
(3) The mass required is $0.05 \times 39.998 \times 5 \times 10^{-3} = 9.999 \times 10^{-3}$ g (to four significant figures).

Example 2
What is the total volume of the solution required given 0.481 g of anhydrous $CuSO_4$ to achieve a concentration of 10 mM? ('Anhydrous' means that there are no associated water molecules bound to the crystal.)

(1) Convert to standard units: concentration $= 1 \times 10^{-2}$ mol l^{-1}.
(2) The molecular mass of $CuSO_4$ is $63.55 + 32.07 + (16.0 \times 4) = 159.62$ g.
(3) The equation

$$\text{mass of chemical} = \text{volume} \times \text{molecular mass} \times \text{concentration}$$

where mass is in grams, volume is in liters, and concentration in moles per liter, rearranges to

$$\text{volume} = \frac{\text{mass of chemical}}{\text{molecular mass} \times \text{concentration}},$$

by dividing both sides of the equation by molecular mass × concentration. (See Topic D1 for more on rearranging equations.) So

$$\text{volume} = \frac{0.481}{159.62 \times (1 \times 10^{-2})} = 0.3013.$$

Thus copper sulfate needs to be added to water and the solution made up to 0.301 3 l.

Other measures of concentration

The measures defined below are of less general use than molarity.

Molality is a measure of solution concentration expressed as moles per kilogram of solvent (mol kg^{-1}), and is a temperature-independent measure of concentration, which is only used occasionally.

Normality: a one normal (1 N) solution contains 1 gram equivalent weight per litre (gEW l^{-1}). An equivalent weight is equal to the molecular weight divided by the valence (replacement H ions or their equivalent). Thus HCl (hydrochloric acid) has a valence of 1, as does NaCl (sodium chloride), whilst in H_2SO_4 (sulfuric acid) the valence is 2, and in phosphoric acid (H_3PO_4) the valence is 3. A gram equivalent weight can regarded as containing 1 mole of 'reactive units'. Normality and equivalent weights are now rather dated terminology, but you may encounter them occasionally.

A useful formula for the amount of solute, m (in grams) required to make v liters of solution of concentration c normal is

$$m = vc \times (\text{gEW}) = vc \times \frac{\text{molecular weight}}{\text{valence}}.$$

Example
How much NaCl must be added to 1.5 l of water to obtain a 2 N solution?

(1) Calculate the molecular weight: $23 + 35.5 = 58.5$ g.
(2) Deduce the valence: valence = 1.
(3) Apply the formula: $m = 1.5 \times 2 \times (58.5/1) = 175.5$ g.

There are also three types of **percentage solutions**: weight/weight, weight/volume and volume/volume. All are parts of solute per 100 total parts of solution. Note that it is **not** per 100 parts solvent.

Weight/weight is also known as percentage composition or %w/w. It is the percent of weight of solute in the total weight of the solution. Percent here is the number of grams of solute in 100 g of solution.

Example
A 10% w/w NaOH solution is made by weighing 10 g NaOH and dissolving in 90 g of water (= 90 ml water, assuming the density of water is 1.0 g ml^{-1}).

Weight/volume is also known as percentage concentration or %w/v. It is the percent of weight of solution in the total volume of solution. Percent here is the number of grams of solute in 100 ml of solution. This is easier to achieve than % w/v as the solute mass can be added to the water and this made up to the known volume using a volumetric flask.

Example
A 10% w/v NaOH solution is made by weighing 10 g NaOH and dissolving in 100 g of solution.

Volume/volume is also known as percentage concentration or %v/v. It is the percent of volume of solute in the total volume of solution % v/v. Percent here is the number of milliliters of solute in 100 ml of solution.

Example
A 10% v/v ethanol solution is 10 ml of ethanol in 100 ml of solution; unless otherwise stated, water is the solvent.

Stock solutions

It is common to make up a concentrated **stock solution** and to prepare other solutions from this. Some laboratory protocols require a range of concentrations, which can be simply achieved by the use of **serial dilutions**.

If you add some solvent to a solution of known concentration and volume you obviously end up with a solution of greater volume and lower concentration. Intuitively, this is a proportional relationship: if you double the volume you will halve the concentration. The key formula is $C_1V_1 = C_2V_2$, where C_1 and C_2 are the initial and final concentrations, respectively; and V_1 and V_2 are the initial and final volumes, respectively.

Usually we want to know how much of the stock solution should be added to how much water (or other solvent) to get the desired concentration and quantity. If we rearrange the equation above we can calculate how much volume of stock solution we require.

Example
If you have a stock solution of 0.01 M (= 0.01 mol l^{-1}) KCl, how do you make up 100 ml of 3 mM (= 3 mmol l^{-1})? It does not matter whether you use SI or non-SI units so much as to make sure you are consistent throughout. Remember to convert prefixes.

- Organize the components. As with any calculation, always be sure that you are using the same units throughout.
  - Initial concentration = 0.01 mol l^{-1}
  - Final volume = 50 ml = 5×10^{-3} l

– Final concentration = 3 mmol l^{-1} = 3×10^{-3} mol l^{-1}

- Apply the formula

$$v = \frac{(3\times10^{-3})\times(5\times10^{-3})}{0.01} = 0.0015,$$

where v is in liters. Thus, the initial volume of stock solution required is 0.0015 l (= 1.5 ml).

- Calculate the volume of water required to add to this:

volume of water required = final volume – initial volume

= 50 – 1.5 = 48.5

- Volume of water required is therefore 48.5 ml.

Serial dilutions

The basic principle of serial dilutions is to start with a stock solution and dilute this by a fixed proportion, then take an aliquot of the diluted solution and dilute this by the same proportion, and so on. There are two common forms: **doubling dilutions** and **decimal dilutions**.

For a doubling dilution, equal volumes of the stock and water are added, this is stirred and then half is transferred to a another vessel, an equal volume of water added, and so on. The concentrations obtained will be 1, $\frac{1}{2}$, $\frac{1}{4}$, $\frac{1}{8}$ etc. multiplied by the original concentration.

For a decimal dilution, measure out one tenth volume of the stock and add nine times as much water, mix thoroughly and repeat. The concentrations obtained will be 1, $\frac{1}{10}$, $\frac{1}{100}$, $\frac{1}{1000}$ etc. times the original concentration.

Exercises

Try these exercises relating to the preparation of chemical solutions. Express your answers as mol l^{-1} or mmol l^{-1}, depending on which is most appropriate.

(1) What is the molarity when
   i. 5.84 g of NaCl is added to 5 l of water?
   ii. 240 g of glucose ($C_6H_{12}O_6$) is added to 1 203 ml of water?

(2) How much rock salt (comprising 83.1% NaCl and 16.9% insoluble impurities) needs to be added to 3 000 l of water to achieve a molarity of 0.55?

(3) One third of a 75 ml batch of 20% w/v zinc nitrate solution is added to 150 ml of distilled water. What is the resulting concentration?

Answers

(1) Determine the molarity from the molecular weight (MW):
   i. MW = 22.99 + 35.45 = 58.44 g,
   5.84 g ≡ 0.1 mole, 0.1 mole in 5 l = 20 mmol l^{-1}.
   ii. MW = $(6 \times 12.011) + (12 \times 1.007) + (6 \times 16)$ = 180.15 g,
   240 g ≡ 1.332 mole, 1.332 mole in 1203 ml = 1.107 mol l^{-1}.

(2) 0.55 mol l^{-1} in 3 000 l = 1 650 moles, which weigh $(1\,650 \times 58.44)$ = 96 426 g. This is 83.1% of the total weight. Thus the total is $(100/83.1) \times 96\,426$ g = 0.116 0 tonne.

(3) Attempts to express this as a molarity may founder on the formula $Zn(NO_3)_2$, MW 189.38. A 20% solution has a molarity of $(1\,000/189.38) \times 20\%$ = 1.056 mol l^{-1}, so 25 ml of this has 0.026 4 mole, which in 175 ml gives 0.150 8 mol l^{-1}, which equals $(1\,000/189.38) \times 2.857\%$.

C1 HANDLING DATA

Key notes

Collecting and collating data	Before you start collecting data it is essential to think through what you wish to show, what data you wish to collect and the best strategy for collating the information. Much time, effort and money can be wasted if data is in a form that is either worthless or hard to interpret.
Summarizing sample data	Descriptive statistics can be used to summarize data and provide estimates of the distribution of the population being sampled.

Collecting and collating data

Before you start collecting data, it is essential to think through the best strategy to collate the information you collect, and to make sure that the best use is made of your efforts. It is surprisingly easy to collect data in a form which is either worthless or hard to interpret, and it is much easier to summarize data for the final report or paper if the raw data is well structured.

Firstly, consider what procedure you are going to undertake with the data. Are you going to undertake any statistical tests? And if not, why not?! (See p. 149.) If so, you need to be sure that the data you collect meets the requirements of the test you wish to use, e.g. replication.

Secondly, it is very useful to construct a pro forma (that is, a ready made template) with spaces for all relevant information. A typical layout is shown in *Table 1*. This will both save time when writing down data and will provide a prompt to ensure that all the necessary information is collected on each occasion.

Summarizing sample data

Biological data very often exhibits variability. In some instances, data will be invariate, such as the number of amino acids in a newly described enzyme, but much more commonly, it will exhibit variation. If a sample exhibits variation,

Table 1. Rocky shore transect data collection sheet (part)

Site:	Transect number:	Transect position:	Date:	Name of recorder:
Species	Quadrat density (individuals m^{-2})			

descriptive statistics can be used to **summarize the data** and **provide estimates of the distribution** of the population being sampled.

By definition a **statistical population** is a set of data that is collected as a representation of the population from which it came (e.g. the population of the lengths of male stag beetle antlers or the population of muscle test responses in 18-year-old footballers).

There are three different features of sample distribution which can be summarized and for which the population values can be estimated:

- **location** — the 'central' value (the mean, median or mode),
- **dispersion** — the spread of the data (variance, standard deviation, standard error and others),
- **skewness** — the symmetry of the frequency distribution.

Measures of location

The **mean** is the most frequently used **measure of location**, and the only one which can be routinely calculated using ordinary pocket calculators. It is not appropriate if the data is ranked or qualitative, or where the data is grouped in classes and may be misleading if the distribution is highly skewed or if there are 'outliers' (anomalous values).It is calculated by summing all the data values and dividing by the sample size (see Box C1 for example of this and other descriptive statistics). The mean is often denoted by $\bar{x}$, $\overline{X}$ or $\overline{Y}$ in statistical texts. Here I will use $\overline{X}$.

The **median** is the **mid-point of observations** when ranked in increasing value. For odd-sized samples, the median is the middle-ranked observation, while for even sized samples, it is the mean of the pair of middle-ranked observations. By definition, half the data values lie above the median and half below it. In a symmetrical distribution the median will be very close or identical to the mean, but with increasing asymmetry the two statistics will deviate as the mean will be affected by the extreme values in the tail of the distribution.

The **mode** is the most common value in a sample and occurs at the **peak of a frequency distribution** plot of the data. There may be more than one modal value. With **continuous data** (see B1 p. 3), it only makes sense to group data into classes and to derive the modal class. The mode is the only measure of location which can be applied to qualitative data (e.g. 'the modal flower color chosen by honey bees was yellow').

Measures of dispersion

The **variance** is the basis of the most commonly used measures of dispersion (although it is not usually used in its own right). It is based on an **averaged difference between each data value and the mean**, so it is zero if all values lie on the mean and is large if many of the values lie far from the mean. The variance is denoted by s^2 and the formula can be expressed as

$$\frac{\text{sum of}\left[(\text{each data value} - \text{mean})^2\right]}{\text{sample size} - 1}$$

or, in mathematical terms:

$$s^2 = \frac{\sum\left(X_i - \overline{X}\right)^2}{n-1}.$$

Standard deviation is simply the square root of the variance. It is denoted by s and its formula can be expressed as

$$s = \sqrt{\text{variance}}$$

or, expressed mathematically,

$$s = \sqrt{\frac{\sum(X_i - \bar{X})^2}{n-1}}.$$

The clear advantage the standard deviation has over the variance is that its units are the same as the variable measured, whilst the variance has units which are the square of these, which is a difficult concept to deal with. For example, a study measuring rabbit population sizes would have variance units of rabbits squared!

Both variance and standard deviation are readily calculated on a pocket calculator (see below). Note that this value is strictly the population standard deviation, and that the sample standard deviation, a value available on most statistical calculators, is of little or no value in everyday life sciences.

Standard error of the mean is the most common way of expressing the variability of data. Every estimated statistic (e.g. mean, variance, skewness) has its own **variability**, and this is known as the **standard error** of that statistic. The standard error of the mean is by far the most commonly used of these, and its name is often shortened to simply 'standard error' (SE). It is calculated by dividing the standard deviation by the square root of the sample size:

$$\text{SE} = \frac{\sqrt{\text{variance}}}{\sqrt{n}} = \frac{\text{standard deviation}}{\sqrt{n}}$$

or, expressed mathematically,

$$\text{SE} = \frac{\sqrt{\frac{\sum(X_i - \bar{X})^2}{n-1}}}{\sqrt{n}}.$$

The **range** is the **difference between the smallest and largest values**. It is of limited value and application, being so strongly affected by variations in chance sampling.

Calculating summative statistics

If you have a calculator with statistical functions, you can efficiently calculate means, standard deviations and other values, without entering the equations above. Different calculators employ slightly different approaches, but the points below are general. A typical scientific calculator is shown in *Fig. 1*.

If you have a **small data set** or **continuous data** not arranged as a frequency table then a simple approach is to

(1) select the stats option on the calculator, if required, then
(2) input each data item in turn – usually as [data value] followed by the **M+** key. If you make an error, the last value input can be removed by the M- key.

Alternatively, if you are inputting a frequency table, this can be done efficiently. For example, you may have a frequency table (e.g. the number of leaf mines found in a sample of holly leaves) which is similar to *Table 2*. There are a total of 128 observations here, but you do not need to enter them one by one.

(1) Select the stats option on the calculator, if required.

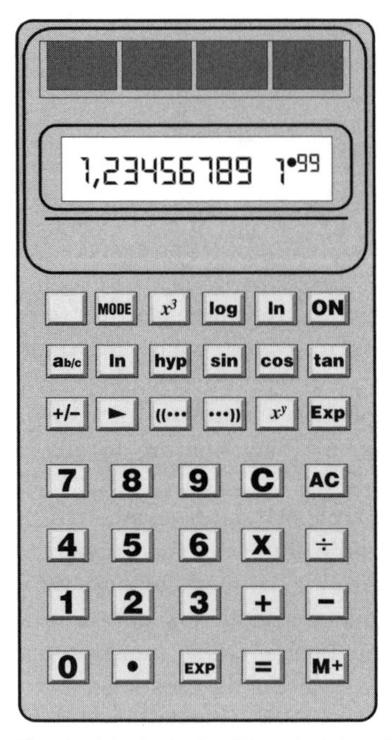

Fig. 1. A typical scientific calculator with statistical functions.

(2) Type [value] [×] (multiplication key) [frequency] M+. For example, for the first line in *Table 2*, you would type

(3) Repeat for each line of the frequency table. Note that you can ignore any lines where the frequency is zero.

Once all values have been input, a range of derived information about the data can be extracted, often including that given in *Table 3*.

Table 2. A typical frequency table

Value	Frequency
0	95
1	25
2	7
3	1
4	0

Table 3. Derived information that can be obtained by using a scientific calculator

Statistic	Meaning
n	Sample size
$\bar{x}$	Mean
σ_{n-1}	Population standard deviation
σ_n	Sample standard deviation
$\sum x$	Sum of data values
$\sum x^2$	Sum of squares of data values

It is important to always use the **population standard deviation**, σ_{n-1} and not the sample standard deviation σ_n. The sample standard deviation is of no value in everyday life science, and its routine presence on calculators only serves to confuse. When sample sizes are large, the two values are quite close, but nevertheless, make sure you always use σ_{n-1} unless explicitly told to do otherwise.

C2 PRESENTING DATA

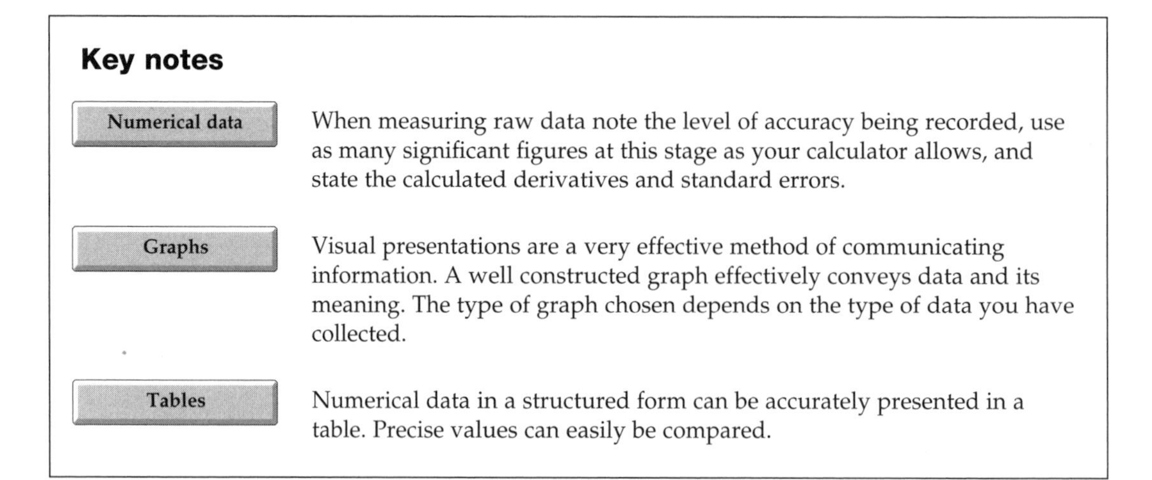

Key notes

Numerical data	When measuring raw data note the level of accuracy being recorded, use as many significant figures at this stage as your calculator allows, and state the calculated derivatives and standard errors.
Graphs	Visual presentations are a very effective method of communicating information. A well constructed graph effectively conveys data and its meaning. The type of graph chosen depends on the type of data you have collected.
Tables	Numerical data in a structured form can be accurately presented in a table. Precise values can easily be compared.

Numerical data

When presenting numerical data, it is important to be clear about the level of **accuracy** which is being described, and not to assign incorrect levels of accuracy to the data values. It may be appropriate to state the accuracy to which numerical results of calculations are given.

As an example of spurious accuracy consider an experiment where seven leaves that were measured to the nearest centimeter had values of 3, 5, 7, 9, 3, 4 and 1 cm. The mean length was calculated and a statement made: 'The mean length of the seven leaves was 4.571428571429 cm.' This statement implies a very high degree of accuracy, and is not justified. The appropriate statement would be: 'The mean length of the seven leaves was 4.6 cm.'

Numerical data should also be stated to the number of **significant figures**. These are the **number of digits which have measurement value**. Zeros with no measurement value are excluded. Thus , the population of the city of Edinburgh to the nearest thousand is 452 000. There are three significant figures in this number. The population of the city of Glasgow to the nearest thousand is 610 000. Note that there are three significant figures in this number too: the first zero conveys information.

Rules about collating and presenting numerical data

(1) When measuring raw data, always write down the data to the level of accuracy being recorded. Thus, if measuring mussel shells to the nearest 0.1 mm using vernier callipers, a shell of 26.31 mm should be recorded as 26.31 mm and not 26.310 mm, while a shell of 24.00 mm should be recorded as 24.00 mm and not 24.0 mm or 24 mm.

(2) When manipulating data in equations or calculating statistics, use as many significant figures as your calculator or computer works to: do not round numbers up or down at this stage.

(3) State calculated derivatives of your raw data (including means) to one more significant figure than the least accurately recorded component of the data was collected to. Thus, if 20 data values were recorded to three significant figures, the calculated mean should be stated to four significant figures. If data comes from several sources to calculate the photosynthetic efficiency of a plant leaf, and some data has been collected to four significant figures, but other data values have been measured only to two significant figures, the overall calculated efficiency value should only be stated to three significant figures.
(4) State standard errors (and related statistics) to one more significant figure than that used for the mean, i.e. two more significant figures than the least accurately recorded component of the data.

Rules 3 and 4 are heuristic (rules of thumb) rather than strict scientific facts. If you follow them you will not go far wrong. However, it may be appropriate on occasions to vary from these. If you are considering very large sample sizes, it may be appropriate to use more significant figures to reflect the reduced sampling error. One could reformulate rule 3 in the following way: if z is the order of magnitude of the sample size n (i.e. $n = 10$, $z = 1$; $n = 100$, $z = 2$; $n = 1000$, $z = 3$; etc.) then calculated values should be stated as z more significant figures than the least accurately recorded component of the data.

Graphs

Visual presentations can be a very effective way of communicating information. A well constructed graph effectively conveys the data and its meaning. In contrast, a table (see below) contains an accurate numerical record of data values. In scientific papers a particular set of data is almost always presented in either graphical or tabular form but not both. In a laboratory practical report, though, you may be asked to express the same data in both formats. The appropriate type of graphical representation you should adopt depends on the type of data you have. Typical plots are shown in *Fig. 1*.

A reminder of some definitions

A **continuous variable** is measured on an uninterrupted scale (e.g. length, mass, temperature).

A **discrete** (or **discontinuous**) variable is measured as counts of different types (e.g. sex, species, color, genotype).

A **dependent variable** has both a 'relaxed' and a 'strict' definition. **Relaxed definition**: any variable hypothesized to depend on another variable, on which it might reasonably be dependent. For example, if butterfly wing muscle activity was recorded during the morning as the air temperature increased, it might be reasonable to treat muscle activity level as a dependent variable, and air temperature as independent. However, in the same study, it would not be reasonable to think that the air temperature was dependent on muscle activity! (See Section K1, p. 149 for why this relaxed definition is statistically problematic and why you should stick to the strict definition where possible.)

Strict definition: any variable measured in a manipulative experiment where another variable (the independent variable) has been controlled, and value(s) of the dependent variable recorded at specific values of the independent variable. For example, (1) the number of *Drosophila* eggs (dependent variable) at different specified temperature values (independent variable); (2) heart rate (dependent variable) at different specified dosages of a pharmaceutical drug (independent variable).

Two continuous variables: both independent

Scatterplot

Two continuous variables: one dependent

Fitted line (linear or curved)

A frequency distribution of a continuous variable

Frequency polygon

(A histogram, a continuous version of the bar chart, may also be used.)

A frequency distribution of a discrete variable

Bar chart

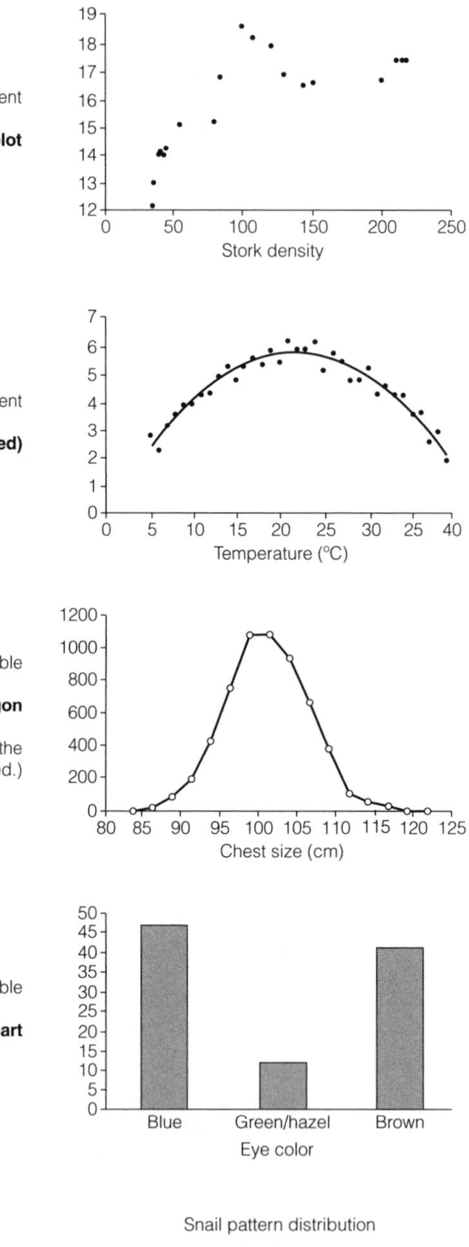

Proportions of a whole

Pie chart

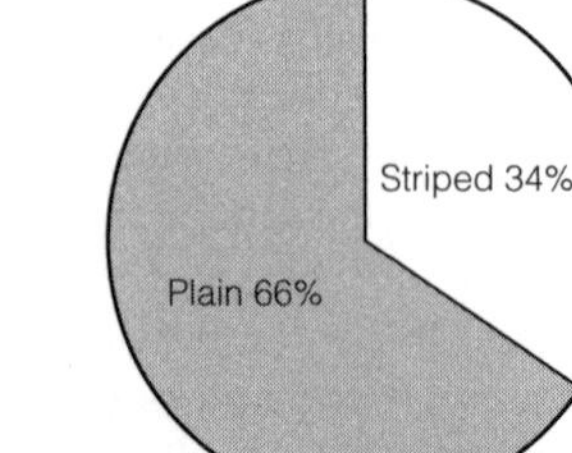

Fig. 1

On a graph, the dependent variable is always plotted on the y axis (the ordinate) and the independent variable on the x axis (the abscissa).

An **independent variable** is not dependent (by one of the above definitions) on any other variable. When plotting two independent variables on a scatter plot, is does not matter which variable is plotted on the x axis and which on the y axis.

Rules for constructing graphs

(1) Choose the appropriate graphical form to depict your data (see above). If you are using computer software to construct your graphs, be aware that software will allow you to make inappropriate choices.

(2) Give the graph a title that is a concise description, e.g. Fig. 2.1 The rate of photosynthesis as a function of temperature at fixed photon flux density. It is not necessary to state 'A graph of …': this is obvious to the reader.

Make sure that wherever appropriate, your graph includes **error bars**, which are usually standard errors (see p. 17) although you may be asked to present other measures of the spread of the data, such as standard deviations or **95% confidence limits**.

Compose a figure legend, including a key for line colors or patterns and symbols. Incorporate statistical values and indicate the nature of the error bars (e.g. bars are ± 1 standard error). (Note that, frequently, the title and the legend are combined.) When complete, look at your graph to check that it can be understood as a free-standing entity.

Tables

A table is a good way of accurately presenting numerical data in a structured form. The data is not so immediately interpretable as a graphical representation, but the precise values can be compared in a way that graphs do not facilitate. However, do not present your data as a table simply because that was the way you collected the raw data!

Rules for constructing a table

(1) Give the table a title that is a concise description of the contents, e.g. Table 3. Tumor responses to selected agents. It is not necessary to state 'A table of …' or 'A table showing …' as it is obvious to the reader that the entity is a table.

(2) Ensure that the correct level of numerical accuracy is used: do not imply spurious levels of accuracy (see p. 4)

(3) Make sure that wherever appropriate, your table includes standard errors (see p. 17) or another measure of the spread of the data as standard deviations or 95% confidence limits (see p. 17 for definitions of these). Cite the number of replicates used in the study.

(4) Wherever appropriate, full details of any statistical tests carried out should be given incorporating statistical values, the replicate size, n, and the probability, p.

(5) Add a footnote if required, to indicate abbreviations and details.

When complete, look at your table to check that it can be understood as a free-standing entity.

D1 MANIPULATING NUMBERS: ALGEBRA

Key Notes

Algebra	Algebra is a simple 'trick' of expressing numerical values and variables as symbols. This is a useful technique as it allows general mathematical patterns to be simply and neatly expressed.
Algebraic expressions	In algebra, equations can be rearranged so that a different component can be the 'subject' of the equation. Data that is presented verbally can be converted to an algebraic form.
Difference and differential equations	Relationships that describe rates of change are expressed by these types of equation.

Algebra

Algebra isn't difficult! The stereotypic image of the wacky wild-haired scientist is incomplete without a blackboard plastered in incomprehensible symbols behind him. However, don't be phased by the notion of converting numbers into letters and other symbols. If you come across an equation or formula which looks troublesome, look at the bits you can understand and go from there. Not knowing what f or a or r means in a given equation does not mean you're daft! If there are symbols or letters that mean nothing to you, don't worry: look for the definitions or ask someone.

In algebra, **symbols** are used to represent two things: **variables** and **constants**. A variable is any one of a collection of numbers. For example we could say that the air temperature in degrees Celsius could be represented by t, or the number of individuals in a population by N. A constant is a fixed value, which may be a precise physical constant like c, the speed of light in a vacuum ($2.997\,924 \times 10^8$ m s^{-1}) or an estimated constant such as the albedo (reflectivity) of a given forest canopy (about 0.17, depending on the study).

When learning mathematics, or when exploring mathematical ideas, it is often convenient to use variables that do not have any specific meaning attached to them. Thus, we might say $y = x + 3$, without defining what x or y might be.

All arithmetic operations which act between numbers also act between variables, as shown in *Table 1*.

Note that

1. the product of a and b is written as ab and not $a \times b$. This avoids confusion between x (the variable) and the multiplication symbol,
2. because of this convention, it is preferable to use single symbols to represent each variable or constant, e.g. t represents temperature, and not to use more letters e.g. *temp*. This avoids ambiguity, as *temp* could be interpreted as the product of t, e, m, and p.

Table 1. Arithmetic operations and their notations

Arithmetic operation		Notation
Sum of *a* and *b*	is written as	$a+b$
Difference of *a* and *b*		$a-b$
Product of *a* and *b*		ab
Quotient of *a* and *b*		a/b or $a \div b$ or $\frac{a}{b}$
Raising *a* to the power *b*		a^b

When dealing with an equation, it is important to remember that the mathematical symbols have particular **priorities** or **precedence rules**. The priority hierarchy is (1) brackets, (2) multiplication and division and (3) addition and subtraction. Thus $3+6\times 2$ equals 15, as it should be read as $3+(6\times 2)$, and not $(3+6)\times 2$ (which equals 18). Equally, $a+bc$ should be read as $a+(bc)$. The meaning of an equation can be clarified by using extra brackets to reinforce these priorities.

Algebraic expressions

One key use of algebra is that equations can be juggled about or rearranged, so that a different component can be the 'subject' of the equation. The **basic rule** of rearranging equations is: **perform the same actions to both sides of the equation**. So if the equation is $6=4+2$, it is possible to perform any manipulation to one side so long as you apply the same manipulation to the other side. For example, if we add 3 to both sides, to give $3+6=4+2+3$, the equality is maintained, just as if we multiply both sides by 2, giving $2\times 6=2\times(4+2)$, or if we divide both sides by 3, $\frac{6}{3}=\frac{4+2}{3}$, or if we multiply both sides by z (an undefined variable): $6z=z(4+2)$.

If we have an equation such as $y=x+2$, and we know the value of y and wish to calculate the value of x, we may wish to rearrange the equation. What we want to do is to get the x on its own on one side of the equation (tradition dictates that this should be on the left) and everything else on the other side of the equation.

This is how we do this:

1. the initial equation is $y=x+2$,
2. subtract 2 from both sides: $y-2=x+2-2$,
3. simplify the equation: $y-2=x$,
4. rewrite the equation: $x=y-2$.

Examples

1. Express the equation $2a=4b+6$ in terms of b.
   i. Subtract 6 from both sides: $2a-6=4b$.
   ii. Divide both sides by 4: $\frac{2a-6}{4}=b$.
   iii. Rearrange: $b=\frac{2a-6}{4}$.
2. Express $u=2v^2-5$ in terms of v.
   i. Add 5 to both sides: $u+5=2v^2$.
   ii. Divide both sides by 2: $\frac{u+5}{2}=v^2$.
   iii. Take the square root of both sides: $\sqrt{\frac{u+5}{2}}=v$.
   iv. Rearrange: $v=\sqrt{\frac{u+5}{2}}$.

3. Express $s = \sqrt{\frac{3t+1}{5}}$ in terms of t.
   i. Square both sides: $s^2 = \frac{3t+1}{5}$.
   ii. Multiply both sides by 5: $5s^2 = 3t + 1$.
   iii. Subtract 1 from both sides: $5s^2 - 1 = 3t$.
   iv. Divide both sides by 3: $\frac{5s^2-1}{3} = t$.
   v. Rearrange: $t = \frac{5s^2-1}{3}$.
4. Express $y = \ln(x) + 2$ in terms of x.
   i. Subtract 2 from both sides: $y - 2 = \ln(x)$.
   ii. Anti-log both sides and raise e to the power of the appropriate expression, which here is $y - 2$: $e^{y-2} = x$.
   iii. Rearrange: $x = e^{y-2}$.

If you follow the examples above, you may be able to spot a pattern: essentially, to transpose the equation (that is, express it in terms of the other variable), you must **reverse each operation in order**. To reverse an addition, you subtract; and so on. *Table 2* gives examples of such inverse operations. (If the examples involving powers or logarithms perplex you, a perusal of Topic D3 may be of benefit.)

As algebra is so useful, you may come across occasions when you want to solve problems by expressing biological or other relationships in terms of algebra. If you are dealing with raw data, it is likely that you will need to statistically analyse the data, for example to detect if significant linear or curvilinear patterns exist (see p. 150). However, sometimes analyzed data is presented verbally rather than algebraically, and it is useful to be able to convert between these two forms.

If, for example, the fitness of the male ground squirrel increases with the number of females he mates with, such that his fitness is equivalent to 2.3 times the number of females mated with, we could express this as $\omega_m = 2.3f$, where ω_m is male fitness (as determined by the total offspring) and f is the number of females mated with. Note that we have had to avoid using 'f' for fitness and 'f' for females, and hence have adopted another symbol, the Greek letter omega, ω. The subscript 'm' is simply added to tell us that this is male fitness; we may also want to construct a similar equation to consider female fitness. The relationship between ω_m and f can be seen by plotting a graph of male fitness (ω_m) as the

Table 2. Operations in algebra and their inverse operations

Operation	Inverse operation
Add	Subtract
Subtract	Add
Multiply	Divide
Divide	Multiply
Square root (equivalent to power $\frac{1}{2}$)	Square (raise to power 2)
Cube root (equivalent to power $\frac{1}{3}$)	Cube (raise to power 3)
Raise to power $\frac{2}{3}$	Raise to power $\frac{3}{2}$
Sine	Arcsine
Cosine	Arccosine
Tangent	Arctangent
Log_e	Raise *e* to power of [expression]
Log_{10}	Raise 10 to power of [expression]
a to power of [expression]	Log_a

ordinate and number of females, f, as the ascissa (*Fig. 1*). (More information about graphs is in Topic C2.)

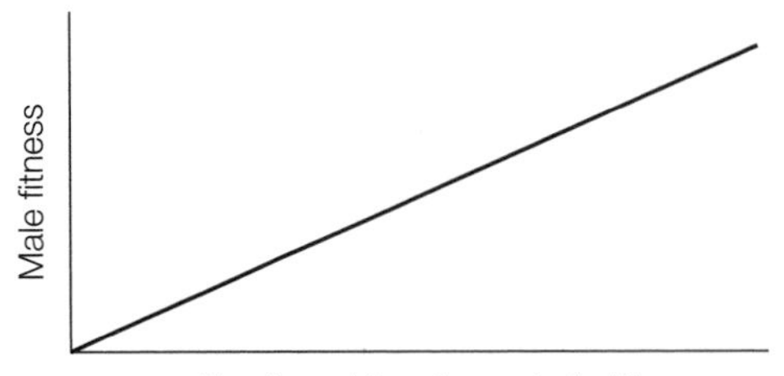

Fig. 1. Reproductive success of the ground squirrel.

Difference and differential equations

There may be more than one way of expressing a relationship algebraically. In particular, if relationships describe rates of change, they can be expressed as either **difference** or **differential** equations. Differential equations are discussed in more detail in Section H, and are instantly recognizable because they start with a term in the form $\frac{dy}{dx}$, where y and x can be any variable.

Athletes possessing the I-form of angiotensin-converting enzyme (ACE) show rapid response to exercise, whilst those possessing the D-form show much slower response. In a standardized arm exercise study, the time to exhaustion decreased by 6.6% per week in individuals homozygous for the I-form and by 0.6% in individuals homozygous for the D-form. These data could be expressed as either:

1. difference equations

$$e_I = e_o - 6.6t,$$
$$e_D = e_o - 0.6t$$

or

2. differential equations

$$\frac{de_I}{dt} = -6.6t,$$

$$\frac{de_D}{dt} = -0.6t,$$

where e_I is time to exhaustion in I-form individuals (in minutes), e_D is time to exhaustion in D-form individuals, e_o is the time to exhaustion in untrained individuals, t is the training period (in weeks). Note that we need to incorporate a starting position (e_o) in the difference equations which is not required in the differential equations.

D2 TRIGONOMETRY

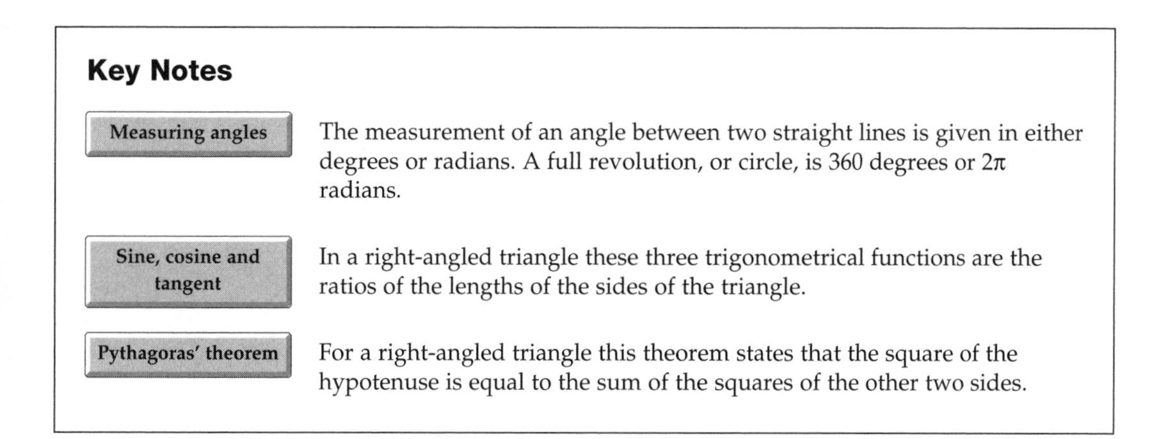

Key Notes

Measuring angles — The measurement of an angle between two straight lines is given in either degrees or radians. A full revolution, or circle, is 360 degrees or 2π radians.

Sine, cosine and tangent — In a right-angled triangle these three trigonometrical functions are the ratios of the lengths of the sides of the triangle.

Pythagoras' theorem — For a right-angled triangle this theorem states that the square of the hypotenuse is equal to the sum of the squares of the other two sides.

Measuring angles

Trigonometry is notionally the study of triangles and the trigonometric functions such as sine, cosine and tangent. It has regular application in physics and engineering, and most life science applications occur near the interface with these disciplines, e.g. in biomechanics.

Before we deal with triangles we will look at the measurement of an angle between two straight lines, such as the angle θ (theta) in the diagram below. This angle is measured by using either **degrees** or **radians** as units.

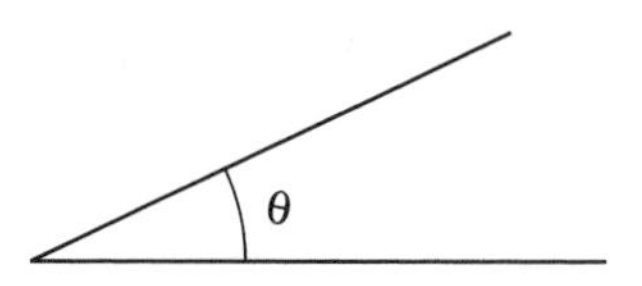

When using degrees, a **full revolution** or circle is **360 degrees** (denoted 360°), and a quarter revolution, or right angle, is 90°. When using radians, a **full revolution** is **2π radians** (where π (pi) is 3.141 59 ...), and a right angle is $\pi/2$. An angle of 1 radian describes an arc of the same length as the radius. It is sometimes particularly useful to use radians rather than degrees as this inbuilt relationship to the geometry can simplify the calculations.

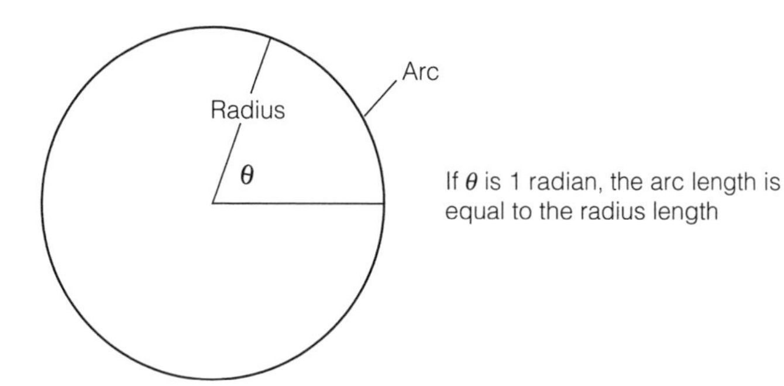

To **convert** an angle expressed in radians to degrees, multiply by $^{360}/_{2\pi}$.

When calculating trigonometric functions, be clear which units (degrees or radians) you are using and avoid mixing them. If it is not apparent to you which units your calculator is set to, calculate sin 90. If the answer is 1, then you are using degrees, but if it is 0.893..., then it is set to radians. Normally a 'Mode' button will select the alternative measure.

Sine, cosine and tangent

If we take a right-angled triangle, the longest side is opposite the right angle and is termed the **hypotenuse**. Taking an angle θ, each of the other sides can be classified as either **opposite** or **adjacent** to θ:

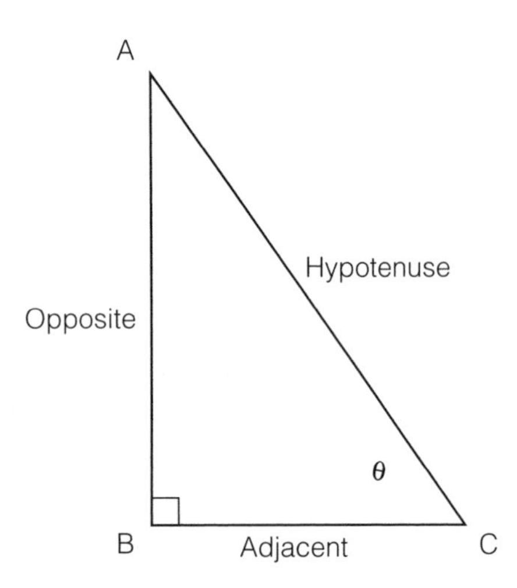

In the triangle ABC shown, side AC is the hypotenuse, side AB is opposite to θ and side BC is adjacent to θ.

If we know the lengths of the sides of the triangle we can calculate the three key trigonometric functions or ratios, sine (abbreviated sin), cosine (cos) and tangent (tan):

$$\sin\theta = \frac{\text{length of side opposite to }\theta}{\text{hypotenuse}} = \frac{AB}{AC},$$

$$\cos\theta = \frac{\text{length of side adjacent to }\theta}{\text{hypotenuse}} = \frac{BC}{AC},$$

$$\tan\theta = \frac{\text{length of side opposite to }\theta}{\text{length of side adjacent to }\theta} = \frac{AB}{BC}.$$

As these ratios are tabulated and built into your calculator, they can be employed to deduce missing information about the triangle, for example if we know the angle θ and the length of the hypotenuse, we can deduce the length of either of the other sides.

Example

Finding an unknown side, AB, of a right-angled triangle, as follows.

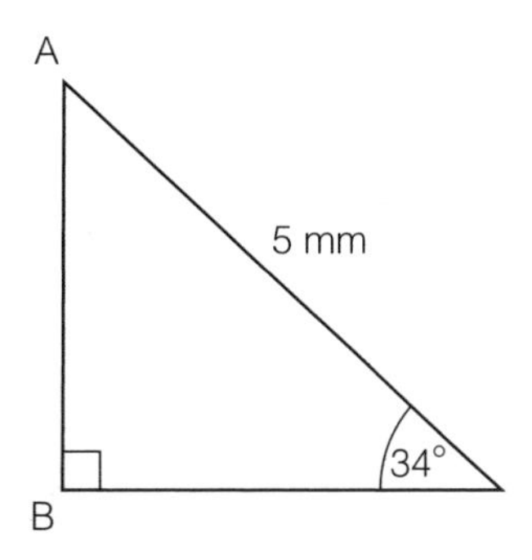

AB is the opposite side to angle θ, and the length of the hypotenuse is 5 mm. So, as $\sin\theta$ = opposite/hypotenuse, then $\sin 34° = {}^{AB}/_{5}$, which rearranges to $AB = 5 \sin 34° = 5 \times 0.559 \ldots = 2.80$ mm.

It is also possible to deduce the angles in a triangle if the lengths of two sides are known.

Example

Finding an unknown angle, θ, in a right-angled triangle.

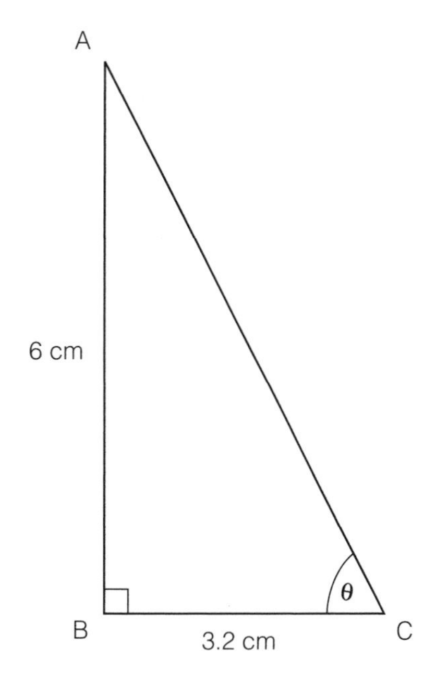

What is the angle θ? We know the lengths of AB, the opposite side and BC, the adjacent side, so, as $\tan\theta$ = opposite/adjacent

$$\tan\theta = \frac{6}{3.2} = 1.875.$$

To find the angle relating to this ratio, we must find the **inverse** function on the calculator, labelled $\tan^{-1}$ (the equivalent functions for sine and cosine are $\sin^{-1}$ and $\cos^{-1}$, respectively), which will return the angle corresponding to the ratio:

$$\tan^{-1} 1.875 = 61.9° = 1.08 \text{ radians}.$$

Pythagoras' theorem

Pythagoras' theorem is named after the Greek mathematician who lived *c*.560–480 BC, but known from a much older Babylonian clay tablet, *c*.1900–1600 BC. The theorem describes the relationship of the length of the sides in a right-angled triangle, and states that **the square of the length of the hypotenuse is equal to the sum of the squares of the other two sides.**

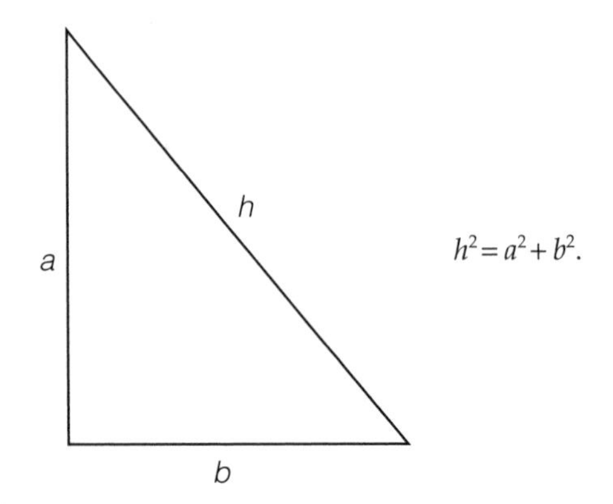

$$h^2 = a^2 + b^2.$$

This relationship allows us to calculate the length of the third side of any right-angled triangle for which we know the length of two sides.

Example
Use Pythagoras' theorem to deduce the length of an unknown side of a right-angled triangle in which the length of the hypotenuse (h) is 8.06 cm and the length of one other side (a) is 4 cm. What is the length of the third side (b)? The equation $h^2 = a^2 + b^2$ rearranges to $b^2 = h^2 - a^2$ (by subtracting a^2 from each side). So, $b^2 = 8.06^2 - 4^2$, and therefore $b = \sqrt{8.06^2 - 4^2} = \sqrt{49} = 7$ cm.

D3 INDICES AND LOGARITHMS

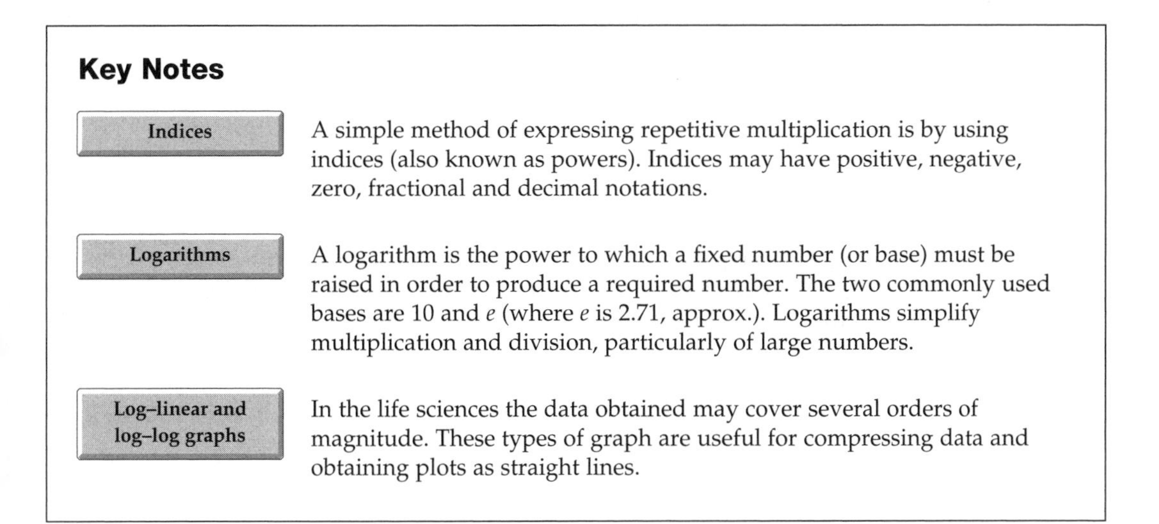

Key Notes

Indices — A simple method of expressing repetitive multiplication is by using indices (also known as powers). Indices may have positive, negative, zero, fractional and decimal notations.

Logarithms — A logarithm is the power to which a fixed number (or base) must be raised in order to produce a required number. The two commonly used bases are 10 and e (where e is 2.71, approx.). Logarithms simplify multiplication and division, particularly of large numbers.

Log–linear and log–log graphs — In the life sciences the data obtained may cover several orders of magnitude. These types of graph are useful for compressing data and obtaining plots as straight lines.

Indices

Multiplication can be defined as an operation which simplifies repetitive addition, as it is much easier to express $4+4+4+4+4+4+4+4$ as 4×8. Similarly, the use of indices (or powers, as they are also known) is a simple way of expressing repetitive multiplication. So, for example,

$$6\times 6 = 6^2 \text{ (described as 6 squared)},$$

$$7\times 7\times 7 = 7^3 \text{ (described as 7 cubed)},$$

$$5\times 5\times 5\times 5 = 5^4 \text{ (described as 5 to the power of 4)}.$$

Indices may be manipulated by **multiplication**, **division** or even as '**powers of powers**'.

As an example of **multiplication**, consider a term such as $4\times 4\times 4\times 4\times 4\times 4$. This can be expressed as 4^6 and also as $(4\times 4)\times(4\times 4\times 4\times 4)$, which is equivalent to $4^2\times 4^4$. In other words, $4^2\times 4^4 = 4^{2+4} = 4^6$. In general, we can state that

$$a^m \times a^n = a^{m+n}.$$

As an example of **division**, consider a term such as

$$\frac{5\times 5\times 5\times 5\times 5}{5\times 5}$$

This is equivalent to

$$\frac{5^5}{5^2}.$$

Cancelling out gives

$$\frac{5\times5\times5\times\cancel{5}\times\cancel{5}}{\cancel{5}\times\cancel{5}}$$

which is equivalent to 5^3. In other words

$$\frac{5^5}{5^2} = 5^{5-2} = 5^3.$$

In general

$$\frac{a^m}{a^n} = a^{m-n}.$$

As an example of '**powers of powers**', a term such as $(3\times3\times3)\times(3\times3\times3)\times(3\times3\times3)$ can be expressed as 3^9, but could also be viewed as $(3^3)^3$. In other words, $3^{3\times3} = 3^9$. In general

$$(a^m)^n = a^{m\times n}.$$

Indices can also be **negative** or **zero**, e.g. 3^{-2} or 3^0. What does this mean? If we consider the progression from positive powers downwards:

$$\begin{aligned}
3^4 &= 3\times3\times3\times3 &&= 81\\
3^3 &= 3\times3\times3 &&= 27\\
3^2 &= 3\times3 &&= 9\\
3^1 &= 3 &&= 3\\
3^0 &= 1 &&= 1\\
3^{-1} &= \tfrac{1}{3} &&= \tfrac{1}{3} = \tfrac{1}{3^1}\\
3^{-2} &= \tfrac{1}{3\times3} &&= \tfrac{1}{3\times3} = \tfrac{1}{9} = \tfrac{1}{3^2}\\
3^{-3} &= \tfrac{1}{3\times3\times3} &&= \tfrac{1}{27} = \tfrac{1}{3^3}
\end{aligned}$$

Note that

(1) 3^1 is 3,
(2) 3^0 is 1,
(3) 3^{-2} is $\frac{1}{3^2}$.

In general

$$\begin{aligned}
a^1 &= a\\
a^0 &= 1\\
a^{-m} &= \tfrac{1}{a^m}.
\end{aligned}$$

The term **fractional indices** refers to notations such as $4^{1/2}$. What does this mean? We already know from the previous information that $a^m \times a^n = a^{m+n}$. So it follows that $4^{1/2}\times4^{1/2} = 4^1 = 4$. In other words, $4^{1/2} = \sqrt{4}$ (= +2 or −2). This is known as the square root of 4. Similarly, $4^{\frac{1}{3}} = \sqrt[3]{4}$ (= 1.587 40 ...), the cube root of 4, and $4^{\frac{1}{4}} = \sqrt[4]{4}$ (= +1.414 21... or −1.414 21...), the fourth root of 4.

Note that the square root of 4 ($\sqrt{4}$) does not have a single solution, because

(1) $+2\times+2 = +4$
and
(2) $-2\times-2 = +4$.

This applies to all even roots (fourth, sixth etc.). Therefore, the solution is written as +2 and −2. Powers such as $5^{2/3}$ can thus be interpreted as $\sqrt[3]{5^2}$.

In general,

$$a^{\frac{1}{m}} = \sqrt[m]{a} \text{ and } a^{\frac{n}{m}} = \sqrt[m]{a^n}.$$

Lastly, you may encounter **decimal indices**. We can evaluate terms such as $4^{1.917}$ or $12^{0.321}$ where the index is not a fraction. In such cases, we can have a heuristic feeling of the likely result (for instance, $4^{1.917}$ is obviously a little less than 4^2 (= 16), and calculation reveals a result of 14.261 (to five significant figures).

It is obvious that it would be extremely tedious to have to calculate indices 'by hand'. Fortunately, many scientific calculators have a specific button or buttons for the most commonly used indices, that is square (x^2) and square root ($x^{½}$ or $\sqrt{x}$) and another which is for all other powers, which is usually marked x^y or y^x.

Examples

(1) To calculate 4^3 use [4] [x^y] [3] [=]
Answer: 64.

(2) To calculate $2.1^{1.5}$ use [2] [•] [1] [x^y] [1] [•] [5] [=]
Answer: 3.0432 (to five significant figures).

(3) Commonly the x^y key also returns the x-th root of y. For $\sqrt[x]{y}$ use shift [ ] [x^y]

(4) To calculate $\sqrt[3]{64}$ use [6] [4] [ ] [x^y] [3] [=]

Logarithms

Logarithms are a way of expressing numbers as powers of a **base**. The two commonly used bases are 10 and e (the natural root) which is 2.718 281 828 (to ten significant figures). If we choose 10, we know that $10^2 = 100$. Therefore the logarithm of 100 to base 10 is 2. Similarly, as $10^3 = 1\,000$, the logarithm of 1 000 to base 10 is 3. We can expand this sequence as shown in *Table 1*.

Note that if $a^n = b$, then $\log_a b = n$. For example, $10^2 = 100$, so $\log_{10} 100 = 2$.

If we say that $\log_{10} 3 = 0.4771$, what does it mean? From the definition above, we know that if $\log_a b = n$, then $a^n = b$. It follows therefore that $10^{0.4771} = 3$. (Check this on a calculator.)

When writing down expressions with logarithms, it is very important to clearly state the base being used. Formally, the base is stated in the way used above as in $\log_{10}$. However, common scientific usage is limited to the bases of 10 and e (the natural root), and the following conventions are used: logarithm to **base 10** is written as **log** or **$\log_{10}$**, and logarithm to **base e** is written as **ln** or **$\log_e$**. When spoken, call this the 'natural log'. If there is likely to be any possibility of confusion, then use the notation $\log_{10}$ or $\log_e$.

Table 1. Numbers and their logarithms

Number	Equivalent to	Logarithm to base 10 ($\log_{10}$)
0.01	10^{-2}	–2
0.1	10^{-1}	–1
1	10^0	0
10	10^1	1
100	10^2	2
1 000	10^3	3
10 000	10^4	4

Just as indices (or powers) can be manipulated, so too can logarithms. Because the logarithm of a number is an exponent or index, the rules which apply to the manipulation of indices translate to logarithms, as shown in *Table 2*.

Table 2. Relationship between functions, indices and logarithms

Function	Indices	Logarithms
Multiplication	$a^m \times a^n = a^{m+n}$	$\log \mathrm{N} + \log \mathrm{M} = \log(\mathrm{NM})$
Division	$\frac{a^m}{a^n} = a^{m-n}$	$\log \mathrm{M} \times \log \mathrm{N} = \log \frac{\mathrm{M}}{\mathrm{N}}$
Exponent	$(a^m)^n = a^{mn}$	$\log \mathrm{M}^\mathrm{N} = \mathrm{N} \log \mathrm{M}$

Before the days of electronic calculators, logarithms used to be part of school mathematics, as their use simplified the multiplication and division of large numbers. As an example consider the calculation of the surface area of Earth, which is given by the formula $4\pi r^2$ (where r, the radius of Earth, is 6 371 000 m).

(1) Convert to logarithms: $\log(4\pi r^2)$.
(2) Since $\log(nm) = \log n + \log m$, then $\log(4\pi r^2) = \log(4\pi) + \log r + \log r$.
(3) By substituting the values for r and π this becomes $\log(4 \times 3.141\,592\,6) + \log 6\,371\,000 + \log 6\,371\,000$.
(4) At this point, in the absence of calculators, it is necessary to refer to logarithm tables, which tabulate values:

$$\begin{array}{r} 1.099\,21 \\ 6.804\,21 \\ 6.804\,21+ \\ \hline 14.707\,63 \\ \hline \end{array}$$

This is the result, expressed in logarithmic form; i.e. the area of Earth's surface is $10^{14.70763}$ m^2. To translate this to normal notation, it is necessary to 'antilog' this number by using logarithm tables again. This gives the surface area of Earth as $5.100\,70 \times 10^{14}$ m^2. (Note that the use of log tables has introduced some rounding error here: the direct calculated value is $5.100\,64 \times 10^{14}$ m^2, to six significant figures.)

Such usage has been superseded by calculators, which could have done the calculation directly, but because logs allow the expression and manipulation of power functions, they are frequently used in professional science applications. Commonly used measures, such as pH (acidity) (see Topic E1) and decibels (sound) are based on logarithmic scales, and logarithmic graphs are commonly used to display data which appear to fit a power function. (See examples of the use of logarithms in the life sciences in Section E.) Thus, logarithms are not out of date! Nevertheless, you are more likely to use a calculator to determine logarithms than to look them up in a set of tables.

Examples

Scientific calculators have a specific buttons for the two most commonly used logarithms – that is to base 10 [log] and to base e (natural log) [ln]

(1) To calculate $\log_{10} 4$ use [log] [4] [=]

Answer: 0.602 06 (to five significant figures).

(2) To calculate $\log_e 3.5$ (i.e. ln 3.5) use [ln] [3] [•] [5] [=]
Answer: 1.252 8 (to five significant figures).

Commonly these keys return the powers of 10^x and e^x, respectively.

(3) To calculate $10^{0.60206}$ use [ ] [log] [0] [•] [6] [0] [2] [0] [6] [=]
Answer: 4.000 0 (to five significant figures). Compare this result to (1), above.

(4) To calculate $e^{1.2528}$ use [ ] [ln] [1] [•] [2] [5] [2] [8] [=]
Answer: 3.5001 (to five significant figures). Compare this result to (2) above. Note that some rounding error has occurred.

Solving problems with logarithms is particularly useful in life sciences, where large populations are often studied. For example, we might wish to know how many generations are required for a single bacterium to create one million offspring, assuming reproduction is by binary fission (*Fig. 1*).

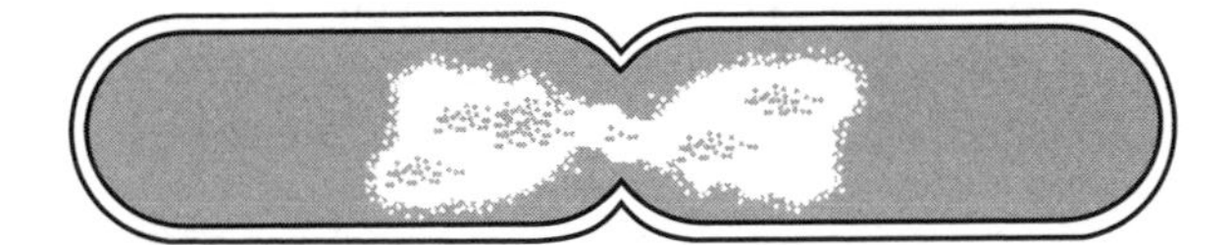

Fig. 1. A dividing bacterial cell.

In the first generation, there will be two offspring; in the second generation, there will be 2^2 (= 4) offspring; in the third generation there will be 2^3 (= 8) offspring, and so on. So we can rephrase the problem as: what is n when $2^n = 1\,000\,000$?

The solution to this problem is

(1) Take logarithms: $\log_{10} 2^n = \log_{10} 1\,000\,000$. (It does not matter which base is used.)
(2) Simplify: $n \log_{10} 2 = \log_{10} 1\,000\,000$.
(3) Get n on the left-hand side: $n = \frac{\log_{10} 1000000}{\log_{10} 2}$.
(4) Evaluate logarithms: $n = \frac{6}{0.30103} = 19.932$ (to five significant figures).

Thus, on the 20th generation, the bacterial population will exceed one million. (*Check*: $2^{20} = 1\,048\,576$.)

In general, logarithms can be used to solve any equation with the general form $x^n = y$, where x and y can be any positive numbers. Note that it does not matter what base is used, although you'd be advised to stick to 10 or e, simply because most calculators will not evaluate logarithms in any other bases.

Log–linear and log–log graphs

It is often useful to plot data on graphs in which one or both scales are logarithmic – that is to say, the unit increments are orders of magnitude (e.g. 1, 10, 100, 1 000). **Log–linear** plots have only one logarithmic axis while in **log–log** plots both axes are logarithmic.

There are two reasons for using log scales:

(1) to compress data collected over a very large range, incorporating several orders of magnitude, use log–log;

(2) if data fits a power function, using the correct plot will result in a straight line.
   i. If the function is of the form $y = ax^b$, use log–log.
   ii. If the function is of the form $y = ae^{bx}$ or $y = a \times 10^{bx}$ or similar, use log–linear.

A straight line on a **log–log** plot means that the underlying function is of the general form $y = ax^b$. The gradient (slope) of this line is exponent b. (It makes no difference whether the base of the logarithm used is 10, e or anything else.) Note that if the underlying function is of the form $y = ax$ (which would give a straight line on an 'ordinary', non-logarithmic, plot) then this will also have a straight line on a log–log plot, of gradient = 1. This makes sense, as $ax^1 = ax$. This is why log–log plots can be used to display linear relationships which cover several orders of magnitude as well as power functions where the exponent does not equal 1.

A straight line on a **log–linear** plot means that the underlying function is of the general form $y = a(\text{base})^{bx}$ where the base can be e or 10, i.e. $y = ae^{bx}$ or $y = a \times 10^{bx}$. The gradient (slope) of this line is exponent b. However, unlike log–log plots it does matter which base you use: if the base is 10, then the relationship for which you are calculating b is $y = a \times 10^{bx}$, while if it is e, the function is $y = ae^{bx}$.

E1 pH, Beer's law, and scaling

Key Notes

The pH scale

pH is a measure of the amount of hydrogen ions in solution: $pH = -\log_{10}[H^+]$. It affects the solubility of many substances and the activity of many biological processes, e.g. intracellular enzyme pathways, behavior of whole organisms.

Absorbance of light

Many biological molecules (e.g. plant pigments) absorb ultraviolet or visible light. The relationship between the amount of radiation absorbed, the wavelength of the light and the concentration of the absorbing species is given by Beer's law, $A = \varepsilon bc$.

Scaling

The relationship between the surface area and volume of an organism can affect key physiological processes. Variations in body weight can lead to significant changes in anatomy and morphology. The dependence of a biological variable on body weight is described by a scaling law, $Y = aM^b$.

Self-thinning

In an even-aged group of sessile organisms, competing individuals cannot escape. The result is fewer individuals of larger size: the process of self-thinning. The relationship between density and individual mass is given by $w = ad^b$. On a log–log graph this leads to Yoda's $-\frac{3}{2}$ law.

The pH scale

pH is a measure of the amount of hydrogen ions (H^+) in a solution, which is a very important measure from a biological point of view: it affects the solubility of many substances, and the activity of many biological processes, from intracellular enzyme pathways to the behaviour of whole organisms. Although it is usual to consider aqueous solutions as containing H^+ ions (protons) in fact they exist as H_3O^+, hydronium ions.

pH is defined as $\mathbf{pH = -\log_{10}[H^+]}$, where $[H^+]$ is the concentration of H^+ ions (in kmol m^3 or the numerically equivalent mol l^{-1}). The preceding minus sign simply makes the pH values positive: as $[H^+]$ values are small, $\log_{10}[H^+]$ is always a negative number.

A **neutral** solution is one in which the concentrations of H^+ ions and OH^- ions are equal. At 25°C, this occurs at a pH of 7.0, but this is temperature dependent, a fact which is important when considering measurements made at different temperatures. Neutral pH is 7.4 at 4°C, but 6.8 at 37°C. The standard range over which pH is normally regarded to lie is between pH = 1 (a very **acidic** solution) where $[H^+]$ is 0.1 mol l^{-1} and pH = 14 (a very **alkaline** solution) where $[H^+]$ is 10^{-14} mol l^{-1}. However, these are not physical limits and whilst pH values outside these limits are uncommon and probably nonexistent in living systems, in theory it is possible to exceed these values.

Because the **pH scale is logarithmic**, a one unit difference is a ten-fold change. A drop in soil pH from 7.4 at one site to 5.2 at an adjacent one represents a 158-fold fall in H^+ concentration. Some textbooks declare that comparative statements such as 'a 0.1 pH unit rise' are wrong because pH is logarithmic. Mathematically, this is untrue as each 0.1 unit change (or any other fixed value) represents the same change in $[H^+]$. For example, a 0.1 rise in pH represents a 126% (or 1.26-fold) rise in $[H^+]$ regardless of the starting value. Nevertheless, from a biological point of view it is probably often valuable to state pH values rather than simply changes.

Absorbance of light

Many molecules in solution absorb ultraviolet or visible light. The absorbance of a solution increases as attenuation of the beam increases, and solutions of higher concentrations exhibit higher absorbance. This is the basis of **colorimetry** and **spectrophotometry**, techniques which allow the concentrations of solutions to be estimated. The amount of radiation absorbed may be measured as either **transmittance** (the proportion or percentage of light passing through) or **absorbance**.

Transmittance, T, may be defined as

$$T = \frac{I_0}{I}$$

where I_0 is the energy of the incident light and I is the energy of the emergent light. Absorbance, A, may be defined as

$$A = \log_{10}\left(\frac{1}{\text{transmittance}}\right) = -\log_{10}(\text{transmittance})$$

or if percentage transmittance is used,

$$A = \log_{10}\left(\frac{100}{\%\ \text{transmittance}}\right) = 2 - \log_{10}(\%\ \text{transmittance}).$$

Thus, the relationship between absorption and transmittance is on a logarithm scale, as shown in *Fig. 1*. If all the light passes through a solution without any absorption, then the absorbance is zero, and the percent transmittance is 100%. If all the light is absorbed, then the percent transmittance is zero, and the absorbance is infinite.

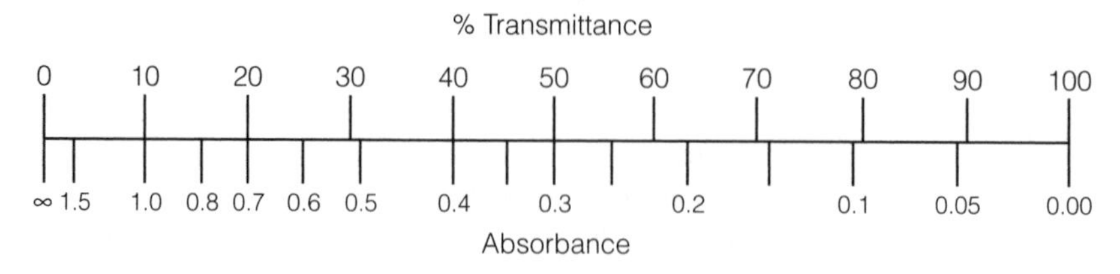

Fig. 1. Relationship between transmittance and absorbance.

Beer's law

The absorbance, A, is directly proportional to the path length, b, and the concentration, c, of the absorbing species according to the relationship

$$A = \varepsilon bc,$$

which is known as **Beer's law**. The absorbance, A, has no units, as it is the logarithm of a ratio. The **molar absorbtivity**, ε, is specific to the molecule being investigated, and has units of $l\ mol^{-1}\ cm^{-1}$ (the number of liters per mole which in a cuvette of 1 cm path length give an absorbance of 1.0). The variable b is the **path length** of the sample (i.e. the path length of the cuvette in which the sample is contained) in cm. The **concentration**, c, is the concentration of the compound in solution, expressed in $mol\ l^{-1}$.

Molecules which are very effective at absorbing light have high molar absorbtivity values (e.g. β-carotene, a protective pigment found in carrots, shrimps and many other species, has a value of $100\,000\ l\ mol^{-1}\ cm^{-1}$, whilst copper sulfate, an 'ordinary' colored solution, has a value of $20\ l\ mol^{-1}\ cm^{-1}$).

You may wonder why Beer's law is expressed in terms of a logarithm (absorbance) and not the simpler transmittance? At first glance it might appear that there is no value in using a logarithm, and that it has just been used to add unnecessary complications! However, this is not the case. If we look at the relationship between the solution concentration and the transmittance and absorbance (*Fig. 2*) we can see that the slope of the transmittance line is constantly changing while that of the absorbance line is constant. In other words, the transmittance/concentration relationship is an exponential decay, and we can convert this into a straight line by expressing the function in terms of absorbance. It is much easier to accurately plot standard known concentrations and estimate the slope of the straight line than to estimate the shape of a curve. To put it another way, if we express Beer's law in terms of transmittance, it becomes $T = \frac{1}{10^{\varepsilon bc}}$, which is much less easy to manipulate than the expression above.

Scaling

Organisms are three-dimensional and the relationships between **length**, **surface area** and **volume** have fundamental impacts on organism **physiology** and **evolution**. If we consider a cubic organism, (unlikely as that may seem), it is easy to see the effect of increasing size. If it has sides of length d then the total surface area is $6d^2$ (six sides each of area $d \times d$) and the volume is d^3. Therefore, as the size of the organism increases, the surface area and volume increase much faster than the length, as shown in *Fig. 3*. This pattern is not substantially altered if the organism is a more usual shape, as the surface area will always be propor-

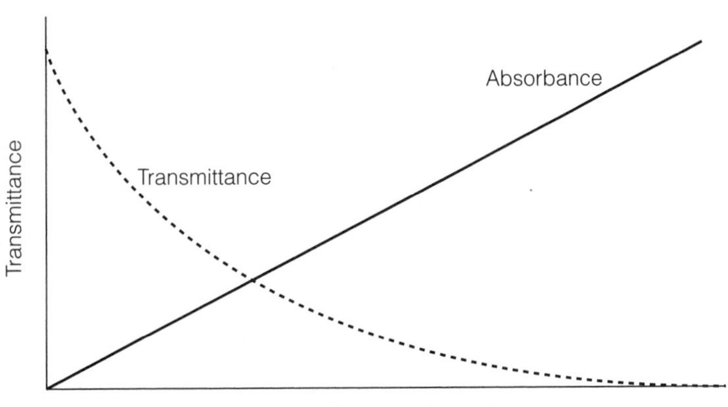

Fig. 2. The relationship between solution concentration, light transmittance and light absorbance.

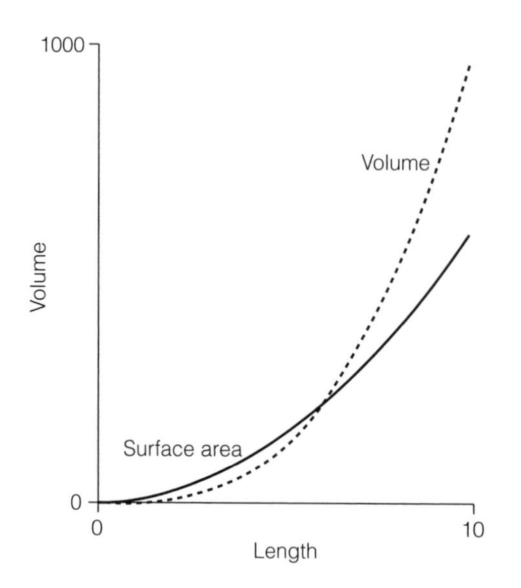

Fig. 3. The changing surface area and volume in a cubic organism as length increases.

tional to the square of the linear dimension, and the volume proportional to the cube of the linear dimension.

Small organisms have a high surface area to volume ratio, but this rapidly changes as the size of the organism increases. This high relative surface area allows small organisms to rely on simple diffusion for some key **physiological processes** such as gas exchange, food uptake, and excretion. Larger organisms are unable to adopt this approach and have evolved complex circulatory systems to ensure tissues are supplied with their needs and have toxins removed. Similarly, as weight is closely related to volume, weight increases cubically as length increases. This means that large animals must have proportionately more skeletal support than small ones. In other words, it is this **cubic scaling relationship** between length and size which means that large land mammals such as elephants and rhinoceroses have fat stumpy legs. It is worth considering that organisms span 21 orders of magnitude in size and a good degree of the variation in **biological diversity** is fundamentally related to this size variation.

As the **weight of an organism** increases, one might expect that **biological variables** such as metabolic rate might increase linearly; after all, more weight must mean more tissue which needs the same per kilogram support. However, the natural world is rather more complicated than this, and many biological parameters increase less rapidly than weight. Early biologists who noted this phenomenon thought that surface area might be the determinant, as it can be reasoned that in endothermic animals, at least, heat loss might be a key determinant of metabolic processes and that this might be driven by surface area. However, this too has proved to be an incomplete answer.

The dependence of a biological variable Y on body weight M can be described by an **allometric scaling law** of the form $Y = aM^b$, where b is the scaling exponent and a is a constant that is characteristic of the kind of organism. If a process is independent of mass, then the exponent b will be zero (as $M^0 = 1$), while if the process increases in proportion to the linear dimension, b will be $\frac{1}{3}$ (as length is proportional to $\sqrt[3]{\text{volume}}$) and if the process increases in proportion to the surface area, b will be $\frac{2}{3}$. Finally, if the process increases proportionally with weight, b will be 1. An example of the dependence of a biological variable (metabolic rate) on the weight of an organism is shown in *Fig. 4*. The upper dotted line indicates the pattern if metabolic rate was proportional to weight ($b = 1$) and the lower dotted line indicates the pattern if metabolic rate was proportional to surface area ($b = \frac{2}{3}$). The observed line lies on a slope of, approximately, $\frac{3}{4}$. Other patterns that we find in the real world are listed in *Table 1*.

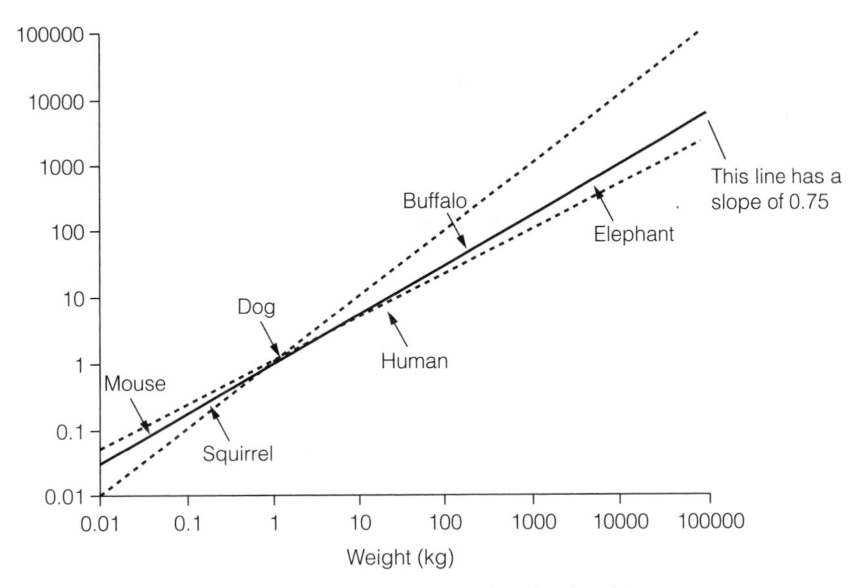

Fig. 4. Metabolic rate in relation to weight in mammals: a log–log plot.

We can understand some of the patterns in the table; for example, we could imagine that blood cells are optimally sized, regardless of the mass of the body they are in, and this may dictate terminal capillary size, while we should be surprised if skeletal mass increased subproportionally to weight. However, the reasons for the prevalence of $\frac{1}{4}$ and $\frac{3}{4}$ power scaling is not wholly understood, and a fulsome discussion of these issues is beyond the scope of this book. Clearly, some process which is general across many species in involved. Many interesting ideas have been put forward to explain these patterns and other related ones (such as the possibility that different allometry rules operate within a species compared to between species). This is another real biological phenomenon which involves powers and for which the use of powers and logarithms is essential.

Self-thinning

In an even-aged group of sessile organisms, competing individuals cannot escape, and typically competition results in the survival of fewer and fewer individuals of larger and larger size. This process is described as **self-thinning**, which results in a relationship between density, d, and individual mass, w, of the form

$$w = ad^b,$$

where a is a constant that is characteristic of the organism. The value of the exponent b is typically close to $-\frac{3}{2}$, a value found to be approximated by

Table 1. Scaling parameters with weight (mostly mammalian)

No change with weight, $b = 0$
Maximum capillary diameter
Red cell size
Mean blood pressure
Fractional airway dead space (VD/VT)
Body core temperature
Maximal tensile strength developed in muscle
Maximum rate of muscle contraction
Mean blood velocity
Scaling to the quarter power, $b = \frac{1}{4}$
Heart rate
Blood circulation time
Respiratory rate
Maximal population growth scale
Embryonic development period
Life-span
Scaling to the three quarter power, $b = \frac{3}{4}$
Most metabolic parameters, including metabolic rate
VO_{2max}
Glucose turnover
Changes in direct proportion to weight, $b = 1$
Heart weight
Lung weight
Tidal volume
Vital capacity
Blood volume
Muscle mass
Skeletal mass

numerous plant species as well as sessile animals like barnacles and mussels. When plotted on a log–log plot, this relationship therefore gives a slope of $-\frac{3}{2}$ (*Fig. 5*). This relationship is known as '**Yoda's law**'. This pattern indicates that in a growing, self-thinning population, weight increases more rapidly than density decreases.

What causes this pattern? Ignoring the finer points of the biological processes that these organisms are involved in, we can consider this as a packing problem. As individual size increases, the surface area occupied will rise as a square of the linear dimension (so density will fall as the inverse of this – a power relationship of $-\frac{1}{2}$) and weight will rise as the cube (i.e. a power relationship of 3). Putting these together results in an exponent of $3 \times (-\frac{1}{2}) = -\frac{3}{2}$.

To illustrate this, consider our cubic organism again and let us assume that it has a density equal to water ($= 1\,000$ kg m^{-3}) and grows from a 'larvae' of 1 mm across to a maximum of 1 m across. In a space of a 1 meter square, 1 000 000 larvae can pack together with a mean weight of 10^{-6} kg. As they grow and self-thinning occurs, the density will fall. When individuals are 1 cm across, the density will be 1 000 m^{-2}; when they have grown to 10 cm across, the density will be 100 m^{-2}; when each surviving individual reaches 50 cm across, the density

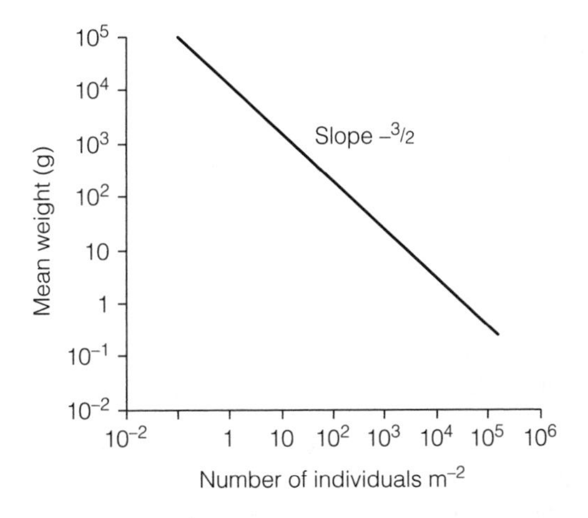

Fig. 5. The relationship between plant density and plant size on a log–log plot, demonstrating Yoda's $-\frac{3}{2}$ law.

will be 4 m^{-2}; and the final point will be reached when one individual fills the 1 meter square plot. At the same time, the mean mass will be rising cubically, from 10^{-6} kg to 10^{-3} kg, 1 kg, 125 kg and, finally, 1 000 kg. The relationship perfectly follows $-\frac{3}{2}$ and thus will give a slope of $-\frac{3}{2}$ on a log–log plot.

The biological feature of interest about such plots is not the approximate gradient of the line (which is inevitable as just described) but the deviations, which are generally small, from this general pattern among different taxa caused by their structural differences.

E2 DEFINING BIOLOGICAL RELATIONSHIPS: FUNCTIONS

Key Notes

Straight lines — A straight line is defined by the equation $y = mx + c$. Different values of x and y will result in lines with different gradients.

Curves — The various types of curve are based on algebraic equations, power functions, and logarithmic functions (both to the base 10 and base e). These are the commonly used functions in life sciences.

Straight lines

It is often useful to relate the relationship of a variable of interest to one or more other factors: the change in metabolic activity on a frog in relation to temperature for example. The simplest kind of relationship is a **straight line**, but sometimes it is necessary to employ a **curved function**, in which case a choice has to be made as to which function is appropriate. It should be noted that whilst the basic mathematics of straight and curved relationships is described here, the process of fitting a function to data is covered by the statistical method of **regression** (see Topic K1).

A straight line can be described by an equation of the form $y = mx + c$, where y is the **dependent variable**, x is the **independent variable**, and m is the **gradient** (or **slope**) and c is the **intercept** of the line with the y-axis (or 'y-intercept'). Some examples of straight-line equations, for various values of x and y are shown in *Fig. 1*.

As any two points can be connected by a straight line it follows that we can define a line in the form $y = mx + c$ from two points on an x–y plot. If the two points are (x_1,y_1) and (x_2,y_2), then the gradient m is given by

$$m = \frac{\text{difference in } y \text{ values}}{\text{difference in } x \text{ values}} = \frac{y_1 - y_2}{x_1 - x_2}$$

The intercept c can then be found by substituting m, x_1 and y_1 into the general equation $y = mx + c$. This can then be rearranged in the form $c = y_1 - mx_1$, and a value for c obtained.

Example

Find the straight line connecting (3,6) and (4,9).

$$m = \frac{y_1 - y_2}{x_1 - x_2} = \frac{6-9}{3-4} = \frac{-3}{-1} = 3,$$

and $c = y_1 - mx_1 = 6 - (3 \times 3) = 6 - 9 = -3$. Thus, the equation is $y = 3x - 3$. As an optional error-check this result can be confirmed by substituting x_2, as the equation should then return y_2:

$$y = 3x - 3,$$
$$y_2 = 3x_2 - 3,$$
$$y_2 = (3 \times 4) - 3 = 12 - 3 = 9.$$

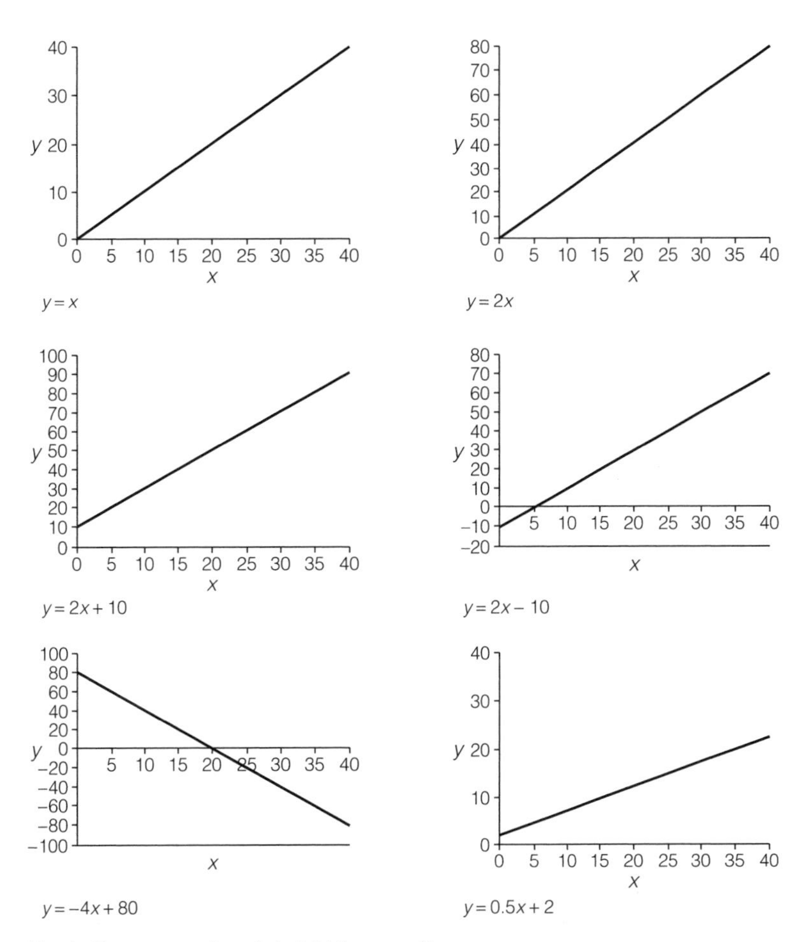

Fig. 1. Some examples of straight-line equations.

Curves

These are most of the commonly used functions in the life sciences. The basic shapes are shown in *Fig. 2*. Note that some functions share similar shapes (such as the sigmoid curve and the logistic growth equation or the logarithmic and power functions), so you should check which is the more appropriate to apply by consulting the following information.

Exponential growth

Equation:

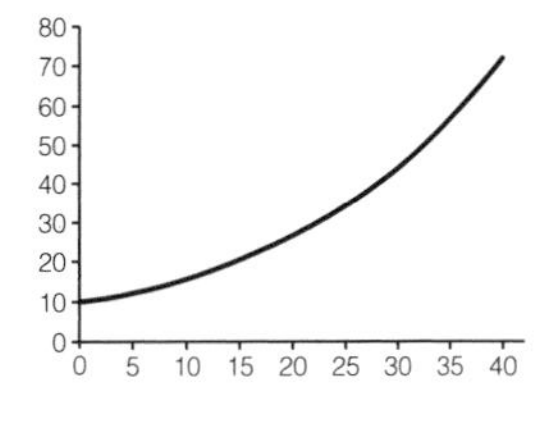

$$y = \text{start} \times e^{bx},$$

where 'start' is the y intercept, b is the growth rate, y is a population size, and x is time.

Use

To describe growth rate of cells, virons, bacteria and other processes under unlimited growth conditions. Exponential growth is also (confusingly) termed 'logarithmic growth', and this terminology is prevalent in microbiology. As resources are finite, exponential growth only occurs for limited periods, whereafter the growth rate slows.

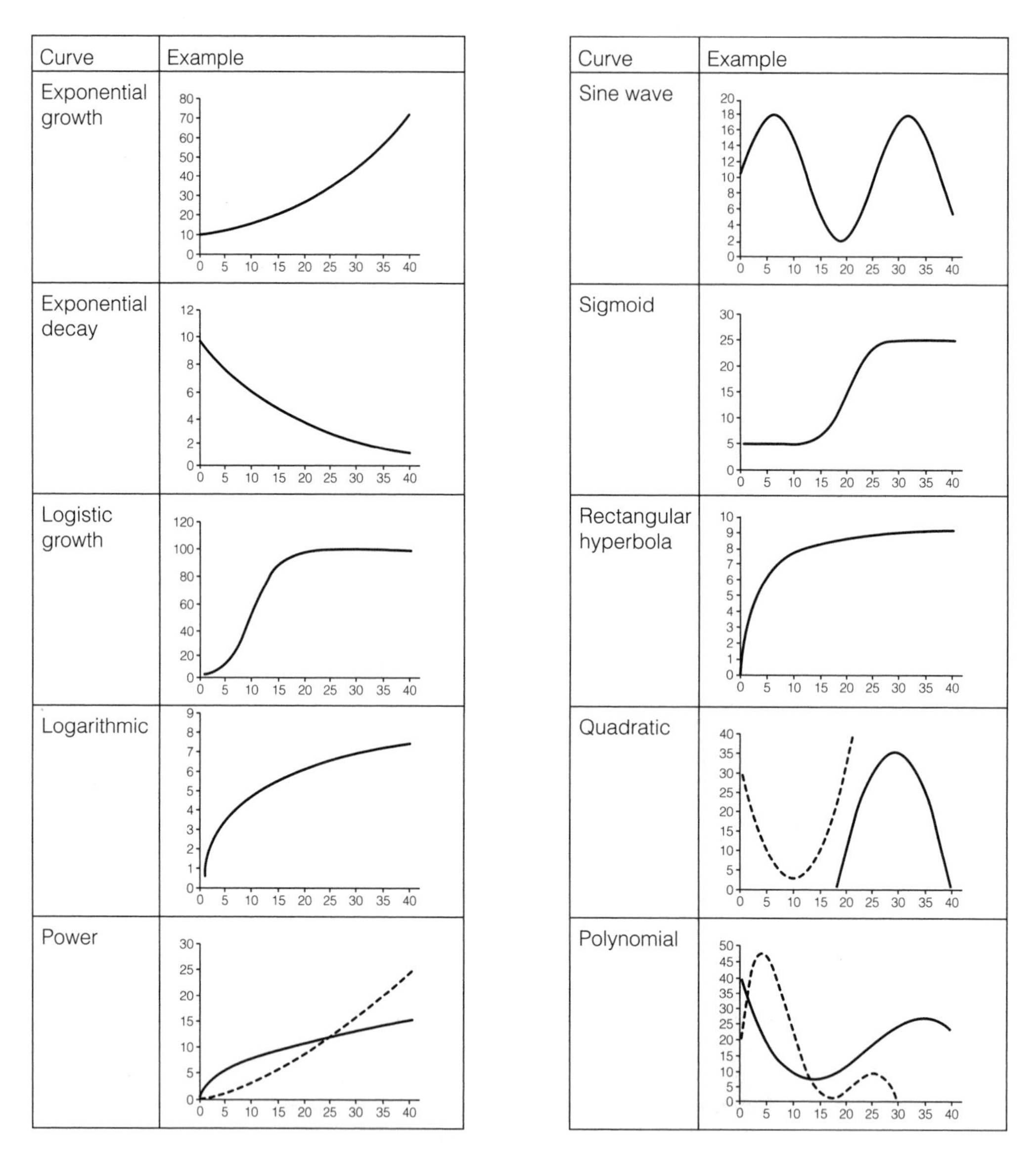

Fig. 2. Examples of commonly used functions in the life sciences.

Exponential decay

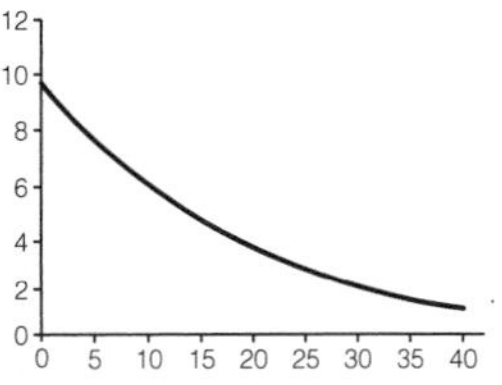

Equation:

$$y = \text{start} \times e^{-bx},$$

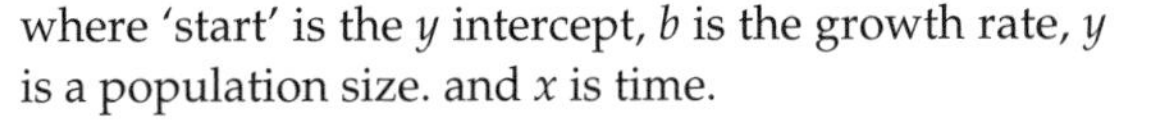

where 'start' is the y intercept, b is the growth rate, y is a population size. and x is time.

Use

To describe the decay of radioactive particles, the loss of nutrients from systems, the decay of drugs in the body and other constant-rate population drops. Cooling bodies also follow an exponential decay towards the ambient environmental temperature.

Radioactivity is often expressed in terms of an element's half-life. For example, the half-life, $t_{\frac{1}{2}}$, of ^{14}C is 5 730 years. This statement means that after 5 730 years half of a sample of, ^{14}C will have undergone decay and become ^{12}C. The half-life is related to the growth rate b as

$$b = \frac{\ln 2}{t_{1/2}}.$$

So, for ^{14}C, $b = \dfrac{\ln 2}{5\,730} \approx \dfrac{0.6931}{5\,730} \approx 1.21 \times 10^{-4}\ \text{year}^{-1}$.

Logistic growth

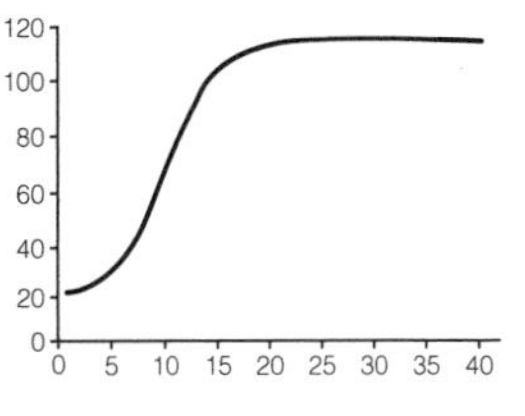

Equation:

$$y = \frac{K}{\left(1 + \left(\frac{K}{N_0} - 1\right)e^{-rx}\right)},$$

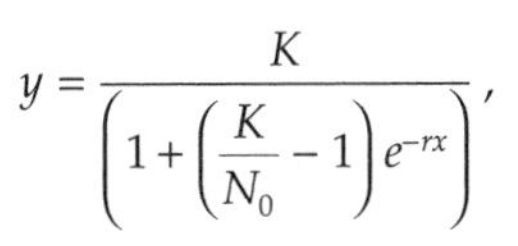

where

y is population size,
x is time,
K is the maximum population size (or 'carrying capacity'),
r is the growth rate (or 'intrinsic rate of increase'),
N_0 is the starting population size at $x = 0$.

Use

This often effectively describes population growth of organisms where resources eventually become limiting.

This equation is often expressed in the difference form, and with time given by T and population size given by N:

$$\frac{dN}{dT} = rN\left(1 - \frac{N}{K}\right).$$

In this form, varying the value of r can result in cyclical and chaotic behavior in the function.

Logarithmic

Equation:

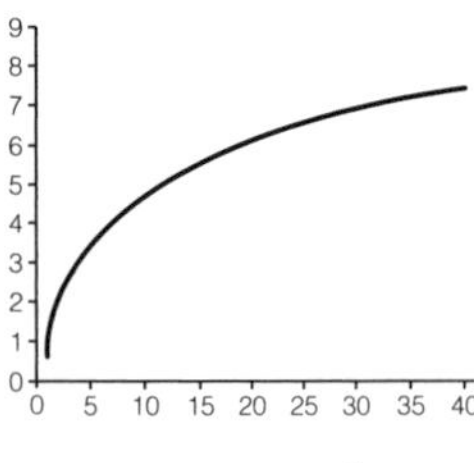

$$y = a\log(bx),$$

where a and b are constants.

Use

This function describes situations where x increases exponentially as y increases linearly. We can rewrite the equation as $bx = e^{y/a}$.

Power

Equation:

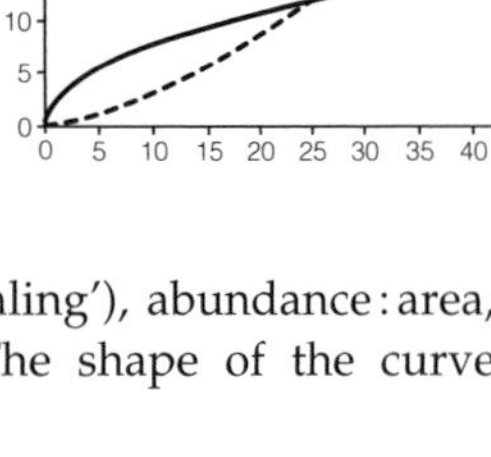

$$y = ax^b,$$

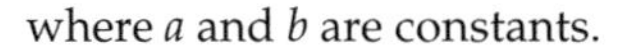

where a and b are constants.

Use

Power relationships often describe proportionalities, such as volume : area, organ mass : body mass (see: 'Scaling'), abundance : area, and abundance at time t : abundance at time $t+1$. The shape of the curve depends on whether the exponent b is above or below 1.

Sine wave

Equation:

$$Y = \text{baseline} + \text{amplitude} \times \sin(fx),$$

where baseline is the y intercept or baseline, amplitude is the amplitude of the cycle (the difference between the baseline and the y maximum (or minimum), f is the frequency of the cycle (where there is one complete cycle every $\frac{2\pi}{f}$ x units, assuming radians are used).

Use

Many regular cyclical patterns can be effectively fitted by a sine wave; for example, diurnal environmental signals such as sunlight.

Sigmoid

Equation:

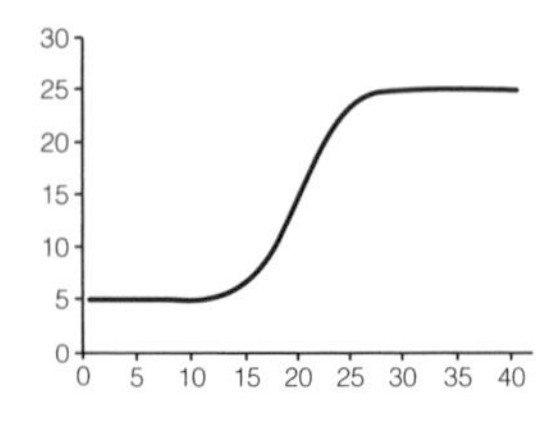

$$y = \text{bottom} + \frac{\text{top} - \text{bottom}}{1 + e^{\text{slope}(\text{center} - x)}},$$

where top is the upper boundary, bottom is the lower boundary, slope describes the slope (positive or negative: a larger value means a steeper slope), center is the x value when y is midway between top and bottom.

Use

Any system where a switch-over from one value to another occurs may be fitted by a sigmoid curve. Examples include voltage activated ion channels (where conductance, y, depends on membrane potential, x), or changes in species presentation or gene frequencies over a linear transect.

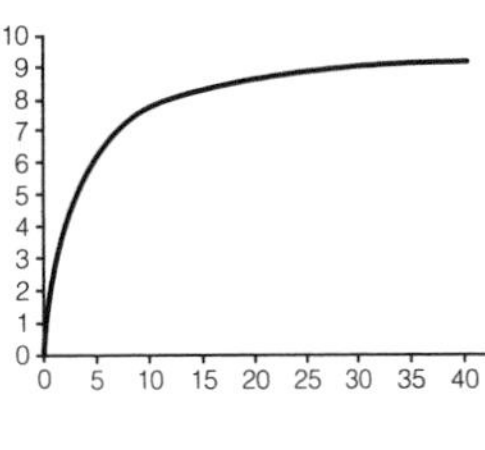

Quadratic
Equation:

$$y = ax + bx^2 + c,$$

where a, b and c are constants.

Use
Can be employed to describe processes with monotonic maxima or minima, such as the response of photosynthesis in a given species to temperature, around an optimum.

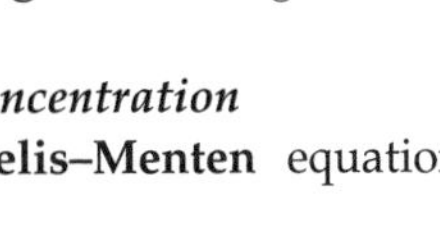

Rectangular hyperbola
Equation:

$$y = \frac{\text{upper} \times x}{\text{center} + x}$$

where upper is the upper boundary and center is the x value when y is $\frac{1}{2} \times$ upper.

Use
This function can describe **enzyme activity** and **ligand binding**.

1. Enzyme activity as a function of substrate concentration
In this case the function is called a **Michaelis–Menten** equation, and is expressed as

$$\text{activity} = \frac{V_{\text{max}} \times \text{concentration}}{K_{\text{m}} + \text{concentration}}$$

or

$$V = \frac{V_{\text{max}} \times [\text{S}]}{K_{\text{m}} + [\text{S}]},$$

where V_{max} is the maximum enzyme activity (or 'velocity') and K_{m} is the Michaelis–Menten constant.

Prior to the availability of nonlinear regression, it was common practice to transform the Michaelis–Menten curve to a straight line. One such approach is the **Lineweaver–Burk** plot (*Fig. 3*), with axes 1/activity (or velocity) and 1/concentration. It is **not recommended** to use a transformed plot to estimate the parameters V_{max} and K_{m} as the transformations distort the experimental error.

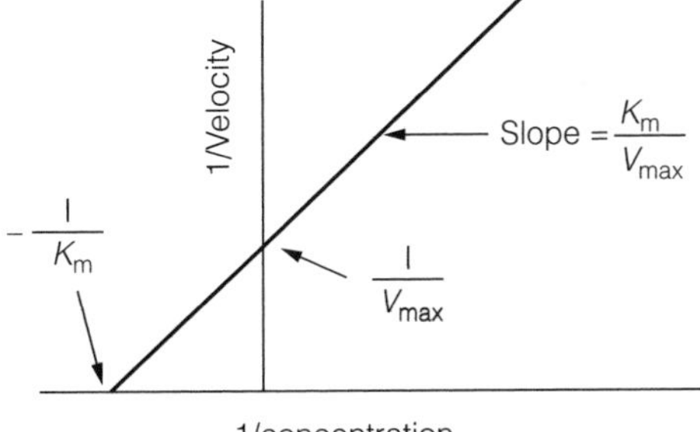

Fig. 3. The Lineweaver–Burk plot.

2. *Ligand binding*

Equilibrium binding of a ligand to a receptor is described, where x is the concentration of the ligand, y is the specific binding, 'upper' is the number of binding sites and 'center' is the equilibrium dissociation constant.

Polynomial

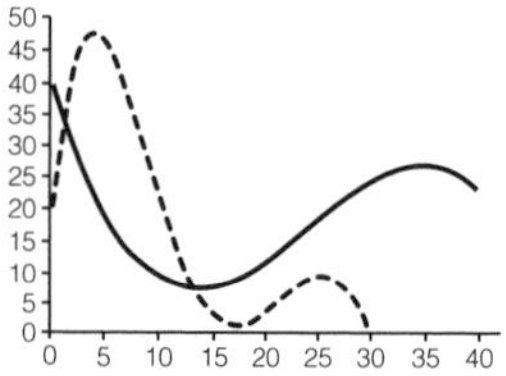

Equation:

$$y = a + bx + cx^2 + dx^3 + \ldots,$$

where a, b, c, d etc. are constants.

There can be as many terms to a polynomial as is desired, although there are a few instances where more than four or five terms are warranted

Use and comments

There are few appropriate applications of polynomials beyond second 'order' (i.e. quadratic, as defined below) that you are likely to come across. Various complex dynamic systems may possess higher order polynomial properties, including a variety of biophysical applications, but you are unlikely to meet them, or have data of sufficient quality to usefully employ such functions. You should be wary of fitting polynomials to noisy data sets, as this technique is widely available but often misused (see warning below).

First, second, and third order equations

Note that the polynomial is an expansion of the quadratic form, which itself can be seen as the expansion of the linear equation. The number of highest power of x in these equations is described as the **order** (or degree) of the equation:

Linear	first order	$y = a + bx$
Quadratic	second order	$y = a + bx + cx^2$
Cubic	third order	$y = a + bx + cx^2 + dx^3$
Quartic	fourth order	$y = a + bx + cx^2 + dx^3 + ex^4$
	n^{th} order	$y = a + bx + cx^2 + dx^3 + ex^4 + \ldots kx^n$

An example of misuse of polynomial line-fitting

Sometimes noisy data, like the simulated set shown in *Fig. 4*, does not offer a good fit to a simple curve or straight line, and it might be tempting to fit a polynomial equation to this, which might appear to be supported by a reasonably good fit. Avoid this temptation!

This 'sixth order' polynomial is apparently a fairly good fit to the data, and indeed the r^2 which describes the proportion of the variation explained by the line (see p. 150) is 83%. However, there is no biological meaningfulness in this pattern, and the line fitted has no predictive value. A linear regression would be the best approach when faced with data like this.

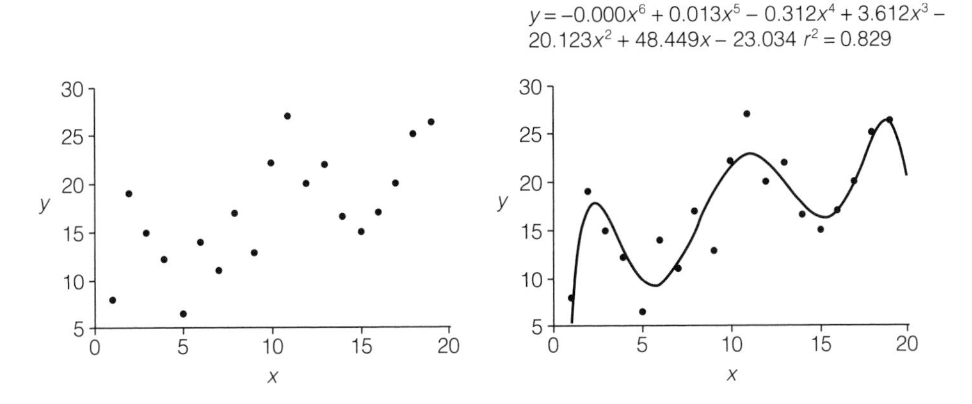

Fig. 4. Simulated examples of 'noisy' biological data.

Exercises

(1) Suppose that for a certain drug, the following results were obtained. Immediately after the drug was administered, the concentration was 6 mg ml^{-1}. Five hours later, the concentration had dropped to 1.2 mg ml^{-1}. Determine the value of b for this drug.

(2) Suppose that for the drug in the previous question, the maximum safe level is 12 mg ml^{-1} and the minimum effective level is 5 mg ml^{-1}. What is the maximum possible time between doses for this drug?

F1 FINDING GRADIENTS AND RATES

Key Notes

Derivative of y with respect to x

If $y = x^2$, the gradient is $2x$. This is the derivative of y with respect to x, and is written as $\frac{dy}{dx}$.

Gradient of a simple function

The gradient of a curve can be found either by drawing the curve on graph paper or differentiating the curve. This is an accurate method and the general rule is: if $y = kx^n$, then $\frac{dy}{dx} = knx^{n-1}$.

Gradient with multiple terms

If y consists of two or more terms added together, then these are differentiated separately and added together: if $y = 2x + x^2$ then $\frac{dy}{dx} = 2 + 2x$.

Derivative of y with respect to x

Variable quantities and their rates of change are dealt with by calculus. There are two related techniques: **differential calculus** (or **differentiation**), which is covered here and **integral calculus** (or **integration**) which is covered in the next chapter.

Differential calculus finds the **gradient** or slope of a function, which is the rate of change of the function at a particular point, such as the instantaneous speed of an accelerating object. If we are interested in the way **non-linear** biological processes such as

- population growth over time,
- enzyme kinetics as the substrate concentration increases,
- the development period of butterfly larvae as the temperature increases

change then we are likely to want to be able to define the rates of change using differentiation. (The rate of change of linear functions is, of course, constant (see Topic D1) though differentiation still works for such functions.) Note that the 'rate of change' does not just apply.

Definitions

If $y = x^2$, the gradient is $2x$. This is the **derivative of y with respect to x**, and is written $\frac{dy}{dx}$. So, if $y = x^2$, $\frac{dy}{dx} = 2x$. More generally, we can find the derivative of any function of the form $y = x^n$ (e.g. $y = x^3$, $y = x^4$, $y = x^{0.5}$) with this **basic rule**:

$$\text{if } y = x^n \text{ then } \frac{dy}{dx} = nx^{n-1}.$$

Example

If $y = x^3$, then $\frac{dy}{dx} = 3x^{(3-1)} = 3x^2$. We can expand this rule to find the derivative of any function of the form $y = kx^n$ (e.g. $y = 3x^3$, $y = 2x^4$, $y = 0.66x^{0.5}$). Thus, the **basic rule expanded** is:

$$\text{if } y = kx^n \text{ then } \tfrac{dy}{dx} = knx^{n-1}.$$

Example
If $y = 2x^3$, then $\frac{dy}{dx} = 2 \times 3x^{(3-1)} = 6x^2$. We can now find the gradient of any function of the form $y = kx^n$.

Gradient of a simple function

As an example, consider the curve $y = x^2$, which describes the pattern of aggregation in the willow-parsnip aphid, where x is the population size and y is the variance in the number of aphids (*Fig. 1*). To find the gradient of this curve at $x = 5$ (or, in other words, to find how rapidly the variance in the number of aphids is increasing at a population size of 5) there are two alternative approaches.

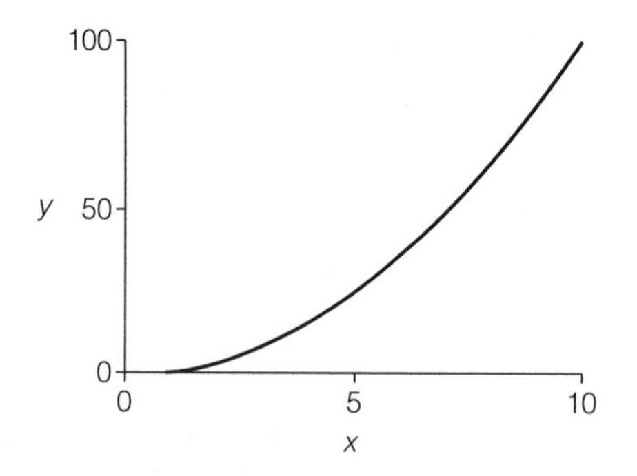

Fig. 1. The pattern of aggregation in the willow-parsnip aphid, where x is the population size and y is the variance in the number of aphids.

(1) Construct the curve on graph paper, and draw a tangent, parallel to the curve, at $x = 5$, and estimate the gradient of this straight line. This method is cumbersome and inaccurate.
(2) Differentiate the curve. This method requires no graphical construction and is accurate.

Apply the general rule: if $y = kx^n$ then $\frac{dy}{dx} = knx^{n-1}$. If $y = x^2$ (so $k = 1$ and $n = 2$) differentiating with respect to x gives

$$\frac{dy}{dx} = 1 \times 2x^{2-1} = 2x.$$

Thus the gradient of the curve $y = x^2$ is $2x$.
At $x = 5$, $\frac{dy}{dx} = 2x = 2 \times 5 = 10$ so the gradient is 10.

Further examples

(1) $y = 2x^3$ $(k = 2, n = 3)$, $\dfrac{dy}{dx} = 2 \times 3x^{3-1} = 6x^2.$

(2) $y = 0.5x^4$ $(k = 0.5, n = 4)$, $\dfrac{dy}{dx} = 4 \times 0.5x^{4-1} = 2x^3.$

(3) $y = 3x^{0.5}$ $(k = 3, n = 0.5)$, $\dfrac{dy}{dx} = 0.5 \times 3x^{0.5-1} = 1.5x^{-0.5}.$

(4) $a = b^3$ $(k = 1, n = 3)$, $\dfrac{da}{db} = 1 \times 3b^{3-1} = 3b^2.$

(5) $y = 6$ $(k = 6, n = 0)^*$, $\frac{da}{db} = 6 \times 0x^{0-1} = 0.$

*Note that we treat '6' as $6x^0$.

Remember that

(1) a variable to the power zero is 1: $x^0 = 1$
(2) a variable to the power 1 is unaltered: $x^1 = x$.

Gradient with multiple terms

If y consists of two or more terms added together, then these are differentiated separately using the basic rules and added together. For example, if $y = 2x + x^2$, then $\frac{dy}{dx} = 2 + 2x$. *Fig. 2* shows the function $y = -0.042x^2 + 1.1x + 9.7$, which describes the effect of temperature in degrees Celsius (x) on the net rate of photosynthesis (y) (in μmol $CO_2\,m^{-2}\,s^{-1}$) of the grass *Deschampsia antarctica*. This function, $y = -0.042x^2 + 1.1x + 9.7$, consists of three additive terms: $-0.042x^2$, $1.1x$ and 9.7. Applying the general rule (i.e. to differentiate kx^n use $\frac{d}{dx} = knx^{n-1}$) to each of these terms and adding the results gives the differentials of the three terms as:

(1) $(k = -0.042, n = 2)$, $\frac{d}{dx} = -0.042 \times 2x^{2-1} = -0.084x.$

(2) $(k = 1.1, n = 1)$, $\frac{d}{dx} = 1.1 \times 1x^{1-1} = 1.1x^0 = 1.1.$

(3) $(k = 9.7, n = 0)^*$, $\frac{d}{dx} = 9.7 \times 0x^{0-1} = 0x^{-1} = 0.$

*Note that we treat '9.7' as $9.7x^0$.

Putting the three terms together, $\frac{dy}{dx} = -0.084x + 1.1 + 0$, which simplifies to $\frac{dy}{dx} = -0.084x + 1.1$.

Further examples

(1) $y = 2x^3 + 6x + 2$, $\frac{dy}{dx} = 6x^2 + 6.$

(2) $y = 3x^3 + 2x^2 + x$, $\frac{dy}{dx} = 9x^2 + 4x + 1.$

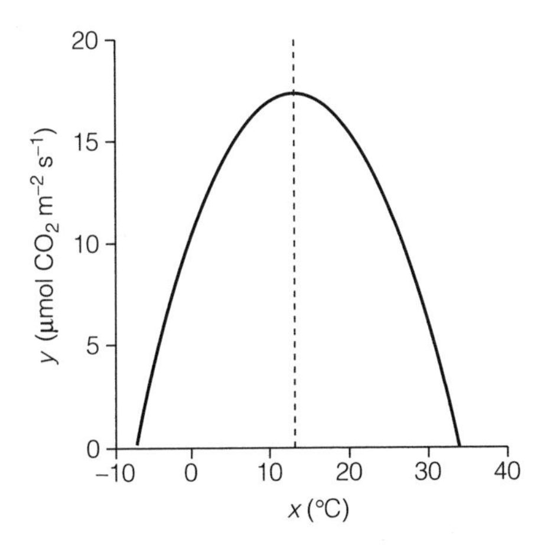

Fig. 2. The photosynthetic response of the grass Deschampsia antarctica *to temperature.*

The above formulae will allow differentiation of simple functions, but not more complex forms (e.g. involving trigonometrical functions, logarithms or complex fractions). It is beyond the scope of this text to consider more complex functions in depth, but it is worth knowing that some of these have simple solutions which are given in *Table 1*.

Table 1. Standard derivatives of functions of x

$y = f(x)$	$\frac{dy}{dx}$
e^x	e^x
e^{ax}	ae^{ax}
a^x	$a^x \ln a$
$\ln x$	$\frac{1}{x}$
$\sin x$	$\cos x$
$\cos x$	$-\sin x$
$\tan x$	$\sec^2 x$

F2 OTHER FUNCTIONS

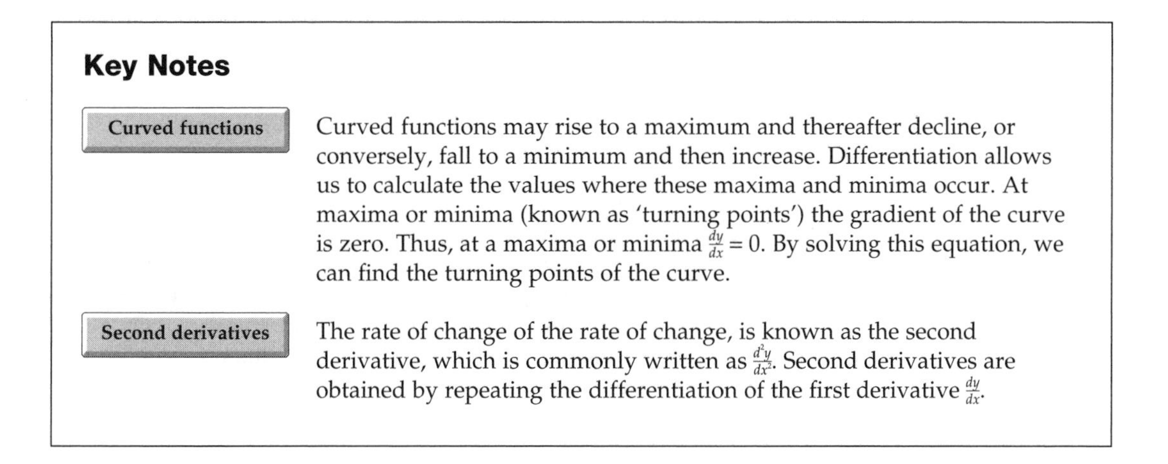

Key Notes

Curved functions — Curved functions may rise to a maximum and thereafter decline, or conversely, fall to a minimum and then increase. Differentiation allows us to calculate the values where these maxima and minima occur. At maxima or minima (known as 'turning points') the gradient of the curve is zero. Thus, at a maxima or minima $\frac{dy}{dx} = 0$. By solving this equation, we can find the turning points of the curve.

Second derivatives — The rate of change of the rate of change, is known as the second derivative, which is commonly written as $\frac{d^2y}{dx^2}$. Second derivatives are obtained by repeating the differentiation of the first derivative $\frac{dy}{dx}$.

Curved functions

Curved functions may rise to a maximum and thereafter decline, or conversely, fall to a minimum and then increase. Such patterns are common in biological systems. For example, enzyme efficiency usually increases with temperature but starts to fall again as the protein is degraded by the heat. Differentiation allows us to calculate the values where these maxima and minima occur. **At maxima or minima** (known as '**turning points**') **the gradient of the curve is zero**. As a curve approaches a maximum, the gradient is positive (like a ball thrown in the air, still rising), whilst after the curve has passed the maxima, the gradient is negative (the ball is falling). At the precise point of the maxima, the gradient is zero (equivalent to the moment the ball is stationary, neither rising nor falling). A similar pattern occurs for a minimum. Thus at a maximum or minimum $\frac{dy}{dx} = 0$. By solving this equation, we can find the turning points of the curve.

If we return to the example of the effect of temperature on the rate of photosynthesis of *Deschampsia antarctica* as shown in Fig. 2 of Topic F1 the graph indicates that the optimum temperature is near 13°C. We can calculate a more accurate value by using the equation $y = -0.042x^2 + 1.1x + 9.7$, therefore $\frac{dy}{dx} = -0.084x + 1.1$. At the turning point $\frac{dy}{dx} = 0$, so $-0.084x + 1.1 = 0$. Subtract 1.1 from both sides to give $-0.084x = -1.1$, then divide both sides by -0.084:

$$x = \frac{-1.1}{-0.084}$$

or $x = 13.1$°C. This means that the maximum photosynthetic efficiency occurs at 13.1°C for the grass *Deschampsia antarctica*.

Second derivatives

As we have seen above, we can find the gradient or rate of change of a curve by differentiating the function. We can take this concept a step further, and find the rate of change of this rate of change. This might seem an odd concept, but it is actually familiar in everyday life. The acceleration of a body is the rate of change of the velocity , which is itself the rate of change of distance with respect to time. Just as we are often interested in the rate of change in biological processes, so we may also be interested in the change of that rate of change.

By convention the second derivative is written $\frac{d^2y}{dx^2}$ (and the third derivative, $\frac{d^3y}{dx^3}$ and so on). The process of calculating a second derivative is simply to repeat the differentiation process. If $y = kx^n$, then, for the **first derivative,** $\frac{dy}{dx} = knx^{n-1}$, and for the **second derivative,** $\frac{d^2y}{dx^2} = kn(n-1)x^{n-2}$. Note that we do not need to employ this second, more complex equation; we can simply differentiate in two steps, repeating the first equation.

Example 1

An object moves away from an observer as shown in *Fig. 1(a)*, with the **distance** (y)/time (x) equation given by $y = 0.011x^3 - 0.7x^2 + 15x - 15$. If this function is differentiated, the rate of change of the distance with respect to time (the **velocity**) is found (*Fig. 1(b)*). If the velocity function is differentiated, the rate of change of the velocity with respect to time is found, i.e. the **acceleration** (*Fig. 1(c)*). A summary of the process is:

Function	Distance/time	$y = 0.011x^3 - 0.7x^2 + 15x - 15$
First derivative	Velocity/time	$\frac{dy}{dx} = 0.033x^2 - 1.4x + 15$
Second derivative	Acceleration/time	$\frac{d^2y}{dx^2} = 0.066x - 1.4$

Example 2

Returning again to the example of the effect of temperature on the rate of photosynthesis of *Deschampsia antarctica*: $y = -0.042x^2 + 1.1x + 9.7$. Therefore $\frac{dy}{dx} = -0.084x + 1.1$ and $\frac{d^2y}{dx^2} = -0.084$. What does this tell us in biological terms? The first derivative is the rate of change of the photosynthetic carbon fixation with respect to temperature. It tells us how fast the carbon fixation rate is changing as temperature changes.

The second derivative is the rate of change of the rate of change of the photosynthetic carbon fixation with respect to temperature. It tells us how fast the rate of change is changing. That the value is constant (i.e. –0.084) tells us that the rate of change of the rate of change of the photosynthetic carbon fixation is constant with respect to temperature.

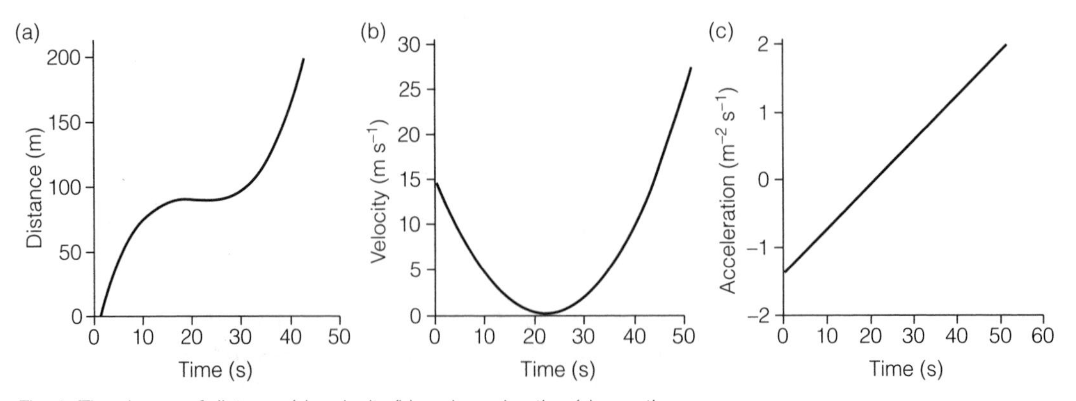

Fig. 1. The change of distance (a), velocity (b) and acceleration (c) over time.

Exercises and answers

Exercise 1

The response of a sense organ to repeated stimulation is described by $y = -0.1x^2 + 6x$, where x is time (seconds) and y is the total number of action potentials generated (*Fig. 2*).

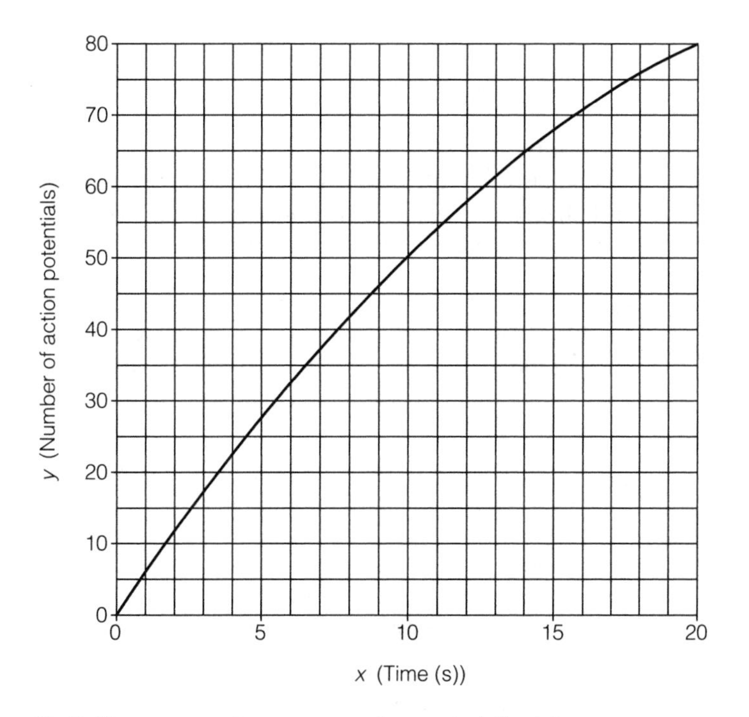

Fig. 2. The response of a sense organ to repeated stimulation.

(1) Estimate the rate of production of action potentials after 13 s (i.e. $x = 13$). Either plot the curve on graph paper or use *Fig. 1* to estimate the slope of a tangent to the curve at $x = 13$. (To calculate the slope of a straight line, see p. 46.)
(2) Differentiate the function $y = -0.1x^2 + 6x$.
(3) What is the value of the slope at $x = 13$ from this result?

Exercise 2

(1) Differentiate the following functions with respect to x (i.e. $\frac{dy}{dx}$), simplifying first where necessary.
 (i) $y = 2x^4 + 2x + 2$
 (ii) $y = 4x^3 + 7$
 (iii) $y = 3x + 3$
 (iv) $y = 4$
 (v) $2y = x^2$
 (vi) $y^2 = 3x$
 (vii) $y = \frac{1}{x}$
 (viii) $y = \frac{1}{2x^3}$

(2) Find the second derivative of
 (i) $y = 2x^4 + 2x + 2$
 (ii) $y = 4x^3 + 7$
 (iii) $y = 3x + 3$

(3) Differentiate the following functions
 (i) $y = na^2$ (with respect to a, i.e. $\frac{dy}{da}$)
 (ii) $s = 3c + 3c^3$ (with respect to c, i.e. $\frac{ds}{dc}$)
 (iii) $d = \frac{1}{2}t^2 + 2t$ (with respect to t, i.e. $\frac{dd}{dt}$)

Exercise 3

The number of individuals, N, in a population of bacteria increases over time, such that $N = rt^2$, where r is a growth factor (and constant) and t is the time after initiation, in hours. What is the rate of change of the population at after 5 h?

Exercise 4

The blood level of sulfanilamide in mice, following injection a the rate of 1 mg per 4 g body weight, is described by $c = -0.77t^2 + 2.39t - 1.06$, where c is $\log_{10}$(sulfanilamide concentration in mg per 100 ml) and t is $\log_{10}$(time after injection, in minutes).

(1) At what time does the maximum blood concentration occur?
(2) What is this maximum concentration?

Exercise 5

An enzyme catalyzed reaction was found to have a velocity, y, which responded to substrate concentration, x, as described by the function $y = -17e^{-33x} + 21$, as shown in *Fig. 3*.

(1) Calculate the differential of this equation, given that the standard differential of ae^{-bx} is $-bae^{-bx}$. For example, if $y = 2e^{-3x}$, then $\frac{dy}{dx} = 6x^{-3x}$.
(2) What is the rate of change of the reaction velocity when the solute concentration is 0.1?

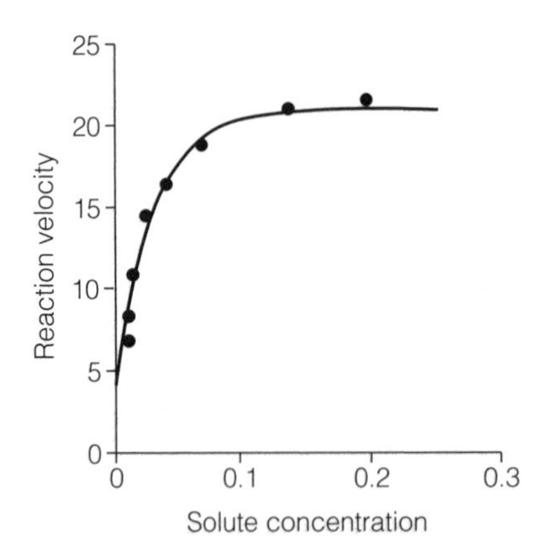

Fig. 3. The relationship between reaction velocity and solute concentration. Points shown are the data. The line shows the fitted function $y = 17e^{-33x} + 21$.

Answers

Answer 1

(1) An estimated tangent at $x = 13$ is shown in *Fig. 4*. The gradient of the tangent, given by $\frac{y_2 - y_1}{x_2 - x_1}$, is thus approximately $\frac{40}{12.5} = 3.2$ action potentials per second. Note that there is nothing special about the size of the triangle chosen to estimate that gradient.

(2) If $y = -0.1x^2 + 6x$ then $\frac{dy}{dx} = -0.2x + 6$.

(3) At $x = 13$, $\frac{dy}{dx} = -0.2x + 6 = 3.4$ action potentials per second.

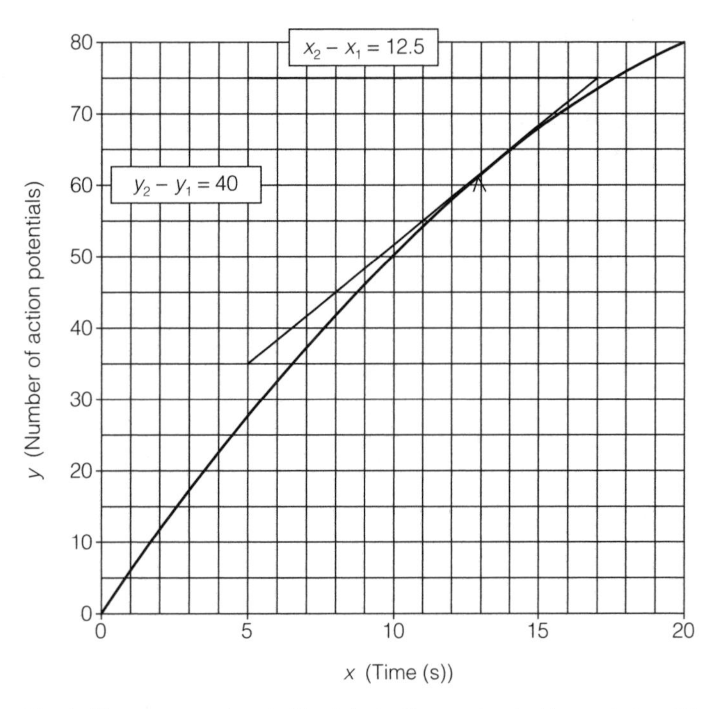

Fig. 4. The same graph as in Fig. 2, but with an estimated tangent at x = 13.

Answer 2

(1) Differentiating functions with respect to x.

(i) $\frac{dy}{dx} = 8x^3 + 2$

(ii) $\frac{dy}{dx} = 12x^2$

(iii) $\frac{dy}{dx} = 3$

(iv) $\frac{dy}{dx} = 0$

(v) First simplify $2y = x^2$ to $y = \frac{1}{2}x^2$ then differentiate: $\frac{dy}{dx} = x$

(vi) First simplify $y^2 = 3x$ to $y = 3^{\frac{1}{2}}x^{\frac{1}{2}}$ then differentiate: $\frac{dy}{dx} = \frac{3^{\frac{1}{2}}}{2}x^{-\frac{1}{2}}$

(vii) $y = \frac{1}{x} = x^{-1}$, so $\frac{dy}{dx} = -x^{-2} = -\frac{1}{x^2}$

(viii) $y = \frac{1}{2x^3} = \frac{1}{2}x^{-3}$ so $\frac{dy}{dx} = -\frac{3}{2}x^{-4} = -\frac{3}{2} \times \frac{1}{x^4}$

(2) Second derivatives

(i) $\frac{d^2y}{dx^2} = 24x^2$

(ii) $\frac{d^2y}{dx^2} = 24x$

(iii) $\frac{d^2y}{dx^2} = 0$

(3) Functions

(i) $\frac{dy}{dx} = 2na$

(ii) $\frac{ds}{dc} = 3 + 9c^2$

(iii) $\frac{dd}{dt} = t + 2$

Answer 3

If $N = rt^2$, then the rate of change of the population over time is given by $\frac{dN}{dt}$, and as $\frac{dN}{dt} = 2rt$ then at $t = 5$ $\frac{dN}{dt} = 10r$.

Answer 4

(1) As $c = -0.77t^2 + 2.39t - 1.06$ then $\frac{dc}{dt} = -1.54t + 2.39$. At the maximum turning point $\frac{dc}{dt} = 0$. Therefore, at this point, $-1.54t + 2.39 = 0$. Hence, $t_{max} = \frac{2.39}{1.54} = 1.551\,9$. As $t = \log_{10}$ (time in minutes), maximum time is therefore 35 min 38.4 s.

(2) $c_{max} = -0.77t_{max}{}^2 + 2.39t_{max} - 1.06$. By substituting $t_{max} = 1.5519$ gives $c_{max} = 0.794\,5$ where c is the $\log_{10}$ (concentration in mg per 100 ml). This gives 6.230 mg per 100 ml.

Answer 5

(1) $\frac{dy}{dx} = 561e^{-33x}$

(2) If $x = 0.1$, $\frac{dy}{dx} = 20.7$

G1 INTEGRATION AND INTEGRALS

Key Notes

Integration as the reverse of differentiation	The process of finding the algebraic form of an expression from its derivative by reversing the process of differentiation is known as integration.
Integrals and standard forms	Integrals of common functions of x are tabulated as standard forms. They are useful when determining indefinite integrals.

Integration as the reverse of differentiation

Integration, as the reverse process of differentiation, is a necessary tool for deriving solutions to differential equations. We shall consider both indefinite and definite integrals as well as numerical integration.

As an example in life sciences, assume that you were told that the rate at which a culture grew on a circular plate was equal to three times the square of the distance from the center of the plate you could write that information as an equation. If we let $A(x)$ represents the amount of culture present at distance x from the center of the plate then the rate at which the culture is growing is represented by the derivative of $A(x)$. That is

$$\frac{dA(x)}{dx}.$$

We are told that this is equal to three times the distance from the center of the plate, so that

$$\frac{dA(x)}{dx} = 3x^2.$$

From this we can calculate the amount $A(x)$ at distance x from the center of the plate by recognizing that if

$$A(x) = x^3 \text{ then } \frac{dA(x)}{dx} = 3x^2.$$

However, this is not the whole story because if, for example, we were take

$$A(x) = x^3 + 5 \text{ then } \frac{dA(x)}{dx} = 3x^2 \text{ as well.}$$

Indeed, there is nothing special about the number 5, we could take any constant and the same thing would happen. That is, if

$$A(x) = x^3 + C, \text{ where } C \text{ is a constant, then } \frac{dA(x)}{dx} = 3x^2.$$

Consequently, we can say that if

$$\frac{dA(x)}{dx} = 3x^2 \text{ then } A(x) = x^3 + C, \text{ where } C \text{ is a constant.}$$

We write this fact using a symbol called the **indefinite integral**:

$$\int 3x^2 dx = x^3 + C.$$

This process of finding the algebraic form of an expression from its derivative by **reversing the process of differentiation** is called **integration** and the arbitrary constant C is called the **integration constant**. You might note that in this example the value of C is, in fact, the amount of culture at the center of the plate. That is, because $A(x) = x^3 + C$ then $A(0) = 0^3 + C = C$, the amount at $x = 0$.

Try some examples. Find each of the following indefinite integrals:

(1) $\int 2x dx$

(2) $\int x^2 dx$

(3) $\int 4x^3 dx$

(4) $\int x^{-2} dx$

(5) $\int \frac{dx}{x^2}$

(6) $\int x^n dx$ where $n \neq -1$

Answers

(1) $\int 2x dx$

Because $\frac{dx^2}{dx} = 2x$ so $\int 2x dx = x^2 + C$.

(2) $\int x^2 dx$

Because $\frac{dx^3}{dx} = 3x^2$ so $\frac{d(x^3/3)}{dx} = x^2$ therefore $\int x^2 dx = \frac{x^3}{3} + C$.

(3) $\int 3x^3 dx$

Because $\frac{dx^4}{dx} = 4x^3$ so $\frac{d(3x^4/4)}{dx} = 3x^3$ therefore $\int 3x^3 dx = \frac{3x^4}{4} + C$.

(4) $\int x^{-2} dx$

Because $\frac{dx^{-1}}{dx} = (-1)x^{-2}$ so $\frac{d(-x^{-1})}{dx} = x^{-2}$ therefore $\int x^{-2} dx = -x^{-1} + C = -\frac{1}{x} + C$.

(5) $\int \frac{dx}{x^2}$

Because $\frac{1}{x^2} = x^{-2}$ so $\int \frac{dx}{x^2} = \int x^{-2} dx = -x^{-1} + C$.

(6) $\int x^n dx$ where $n \neq -1$

Because $\frac{dx^{n+1}}{dx} = (n+1)x^n$ so $\frac{d}{dx}\frac{\left(\frac{x^{n+1}}{n+1}\right)}{dx} = x^n$ therefore $\int x^n dx = \frac{x^{n+1}}{n+1} + C$

provided $n \neq -1$.

Integrals and standard forms

Standard forms

We have just seen in the last section that we can integrate x to any power other than the power -1 by using the simple formula (check this by differentiating the right hand side):

$$\int x^n dx = \frac{x^{n+1}}{n+1} + C \text{ provided } n \neq -1$$

This is an example of what is called a **standard form** and it stands in the same place as the standard forms do in differential calculus. We shall look at some other standard forms later but for now we shall consider the integrals of sums and differences.

Sums and differences

The integral of a sum is equal to the sum of the integrals. For example

$$\int (x^2 + x^3)dx = \int x^2 dx + \int x^3 dx = \frac{x^3}{3} + C_1 + \frac{x^4}{4} + C_2.$$

We write this as

$$\int (x^2 + x^3)dx = \int x^2 dx + \int x^3 dx = \frac{x^3}{3} + \frac{x^4}{4} + C,$$

where $C = C_1 + C_2$, because the sum of two arbitrary constants is still an arbitrary constant.

Similarly for differences, for example

$$\int (x - x^4)dx = \int x dx - \int x^4 dx = \left(\frac{x^2}{2} + C_1\right) - \left(\frac{x^5}{5} + C_2\right) = \frac{x^2}{2} - \frac{x^5}{5} + C,$$

where $C = C_1 - C_2$. Again, the difference of two arbitrary constants is still an arbitrary constant.

So find each of the following integrals:

(1) $\int (x^2 - x)dx$

(2) $\int (x^3 + x^5)dx$

Answers

(1) $\int (x^2 - x)dx = \int x^2 dx - \int x dx = \frac{x^3}{3} + C_1 - \left(\frac{x^2}{2} + C_2\right) = \frac{x^3}{3} - \frac{x^2}{2} + C,$
where $C = C_1 - C_2$.

(2) $\int (x^3 + x^5)dx = \int x^3 dx + \int x^5 dx = \frac{x^4}{4} + C_1 + \frac{x^6}{6} + C_2 = \frac{x^4}{4} + \frac{x^6}{6} + C,$
where $C = C_1 + C_2$.

Scalar multiples

Scalar multiples can be brought outside the integral. For example

$$\int 4x^2 dx = 4\int x^2 dx = 4\left(\frac{x^3}{3} + C_1\right) = \frac{4}{3}x^3 + 4C_1 = \frac{4}{3}x^3 + C,$$

where $C = 4C_1$, because an arbitrary constant multiplied by a scalar is still an arbitrary constant. So try each of these:

(1) $\int 2x^5 dx$

(2) $\int \frac{x}{4} dx$

(3) $\int (2x^3 + 3x^5)dx$

(4) $\int(3x^2 - 4x)dx$

Answers

(1) $\int 2x^5 dx = 2\int x^5 dx = 2\left(\frac{x^6}{6} + C_1\right) = \frac{x^6}{3} + 2C_1 = \frac{x^6}{3} + C$

(2) $\int \frac{x}{4} dx = \frac{1}{4}\int x dx = \frac{1}{4}\left(\frac{x^2}{2} + C_1\right) = \frac{x^2}{8} + \frac{C_1}{4} = \frac{x^2}{8} + C$

(3)
$$\begin{aligned}
\int(2x^3 + 3x^5)dx &= \int 2x^3 dx + \int 3x^5 dx \\
&= 2\int x^3 dx + 3\int x^5 dx \\
&= 2\left(\frac{x^4}{4} + C_1\right) + 3\left(\frac{x^6}{6} + C_2\right) \\
&= \frac{x^4}{2} + 2C_1 + \frac{x^6}{2} + 3C_2 = \frac{x^4}{2} + \frac{x^6}{2} + C
\end{aligned}$$

(4)
$$\begin{aligned}
\int(3x^2 - 4x)dx &= \int 3x^2 dx - \int 4x dx \\
&= 3\int x^2 dx - 4\int x dx \\
&= 3\left(\frac{x^3}{3} + C_1\right) - 4\left(\frac{x^2}{2} + C_2\right) \\
&= x^3 + 3C_1 - (2x^2 + 4C_2) \\
&= x^3 - 2x^2 + 3C_1 - 4C_2 = x^3 - 2x^2 + C.
\end{aligned}$$

So far we have only integrated simple expressions involving powers of x. We have seen earlier that we can differentiate other expressions involving x, for example

$$\frac{d \sin x}{dx} = \cos x \text{ and so } \int \cos x dx = \sin x + C.$$

Similarly

$$\frac{de^x}{dx} = e^x \text{ and so } \int e^x dx = e^x + C.$$

Make sure that you do not forget the integration constant. By proceeding in this manner we can construct a table of standard integrals – called a table of **standard forms**, as shown in *Table 1*. For example, to find $\int(3x^2 - 4\sin x + 5e^x)dx$ we use the table and the properties of the indefinite integral thus:

$$\begin{aligned}
\int(3x^2 - 4\sin x + 5e^x)dx &= 3\int x^2 dx - 4\int \sin x dx + 5\int e^x dx \\
&= 3\left(\frac{x^3}{3} + C_1\right) - 4(-\cos x + C_2) + 5(e^x + C_3) \\
&= (x^3 + 3C_1) - (-4\cos x + 4C_2) + (5e^x + 5C_3) \\
&= x^3 + 4\cos x + 5e^x + 3C_1 - 4C_2 + 5C_3 \\
&= x^3 + 4\cos x + 5e^x + C \text{ where } C = 3C_1 - 4C_2 + 5C_3.
\end{aligned}$$

So try these:

(1) $\int(2x^3 + 3\cos x - e^x)dx$

(2) $\int\left(\frac{2}{x} + \frac{x}{2}\right)dx$

Table 1. Standard integrals (standard forms) for various functions of x

Function	Standard integral
$f(x)$	$\int f(x)\,dx$
x^n	$\frac{x^{n+1}}{n+1}+C \quad n \neq -1$
x^{-1}	$\ln x + C$
$\sin x$	$-\cos x + C$
$\cos x$	$\sin x + C$
e^x	$e^x + C$

(3) $\int(2\cos x - 3\sin x)dx$

(4) $\int(x^3 + 2x)dx$

(5) $\int(2x^2 - 3x^3)dx$

(6) $\int \frac{df(x)}{dx}dx$ given that $\frac{df(x)}{dx} = 5x^3 - 2x^2$ and that $f(0) = 7$

(7) $\int \frac{dg(x)}{dx}dx$ given that $\frac{dg(x)}{dx} = 4x^4 + 3x^2 - 2$ and that $g(1) = 2$

Answers

(1) $$\begin{aligned}\int(2x^3 + 3\cos x - e^x)\,dx &= 2\int x^3dx + 3\int \cos x dx - \int e^x dx \\ &= 2\frac{x^4}{4} + C_1 + 3\sin x + C_2 - e^x - C_3 \\ &= \frac{x^4}{2} + 3\sin x - e^x + C.\end{aligned}$$

(2) $$\begin{aligned}\int\left(\frac{2}{x} + \frac{x}{2}\right)dx &= 2\int\frac{dx}{x} + \frac{1}{2}\int xdx \\ &= 2\ln x + C_1 + \frac{1}{2}\times\frac{x^2}{2} + C_2 \\ &= 2\ln x + \frac{x^2}{4} + C.\end{aligned}$$

(3) $$\begin{aligned}\int(2\cos x - 3\sin x)dx &= 2\int \cos x dx - 3\int \sin x dx \\ &= 2\sin x + C_1 - 3(-\cos x + C_2) \\ &= 2\sin x + 3\cos x + C.\end{aligned}$$

(4) $$\begin{aligned}\int(x^3 + 2x)dx &= \int x^3dx + 2\int xdx \\ &= \frac{x^4}{4} + x^2 + C\end{aligned}$$

(5) $$\begin{aligned}\int(2x^2 - 3x^3)dx &= 2\int x^2dx - 3\int x^3dx \\ &= \frac{2x^3}{3} - \frac{3x^4}{4} + C\end{aligned}$$

(6) $\int \frac{df(x)}{dx} dx = \int (5x^3 - 2x^2) dx$ and that $f(0) = 7$

$= 5\int x^3 dx - 2\int x^2 dx$

$= \frac{5x^4}{4} - \frac{2x^3}{3} + C$ that is $f(x)$ and since $f(0) = 7$ then $0 - 0 + C = 7$

giving the final answer as $\frac{5x^4}{4} - \frac{2x^3}{3} + 7$.

(7) $\int \frac{dg(x)}{dx} dx = \int (4x^4 + 3x^2 - 2) dx$

$= 4\int x^4 dx + 3\int x^2 dx - 2\int dx$

$= \frac{4x^5}{5} + x^3 - 2x + C$ this is $g(x)$ and since $g(1) = 2$ then

$\frac{4}{5} + 1 - 2 + C = 2$ so $C = \frac{11}{5}$ giving the final answer as $\frac{4x^5}{5} + x^3 - 2x + \frac{11}{5}$.

G2 Position, velocity and acceleration

Key notes

Position

The position (i.e. location) of a moving body at any time can be described mathematically by a differential equation.

Velocity and acceleration

Velocity is the rate of change of distance. Acceleration is the rate of change of velocity. Both can be described mathematically by differential equations.

Position

If we denote by $x(t)$ the position x of a moving body at time t then the rate of change of position is the velocity $v(t)$ of the body at time t. That is

$$v(t) = \frac{dx(t)}{dt}.$$

For example, if $x(t) = 3\sin 2t$ then $v(t) = \frac{dx(t)}{dt} = \frac{d3\sin 2t}{dt} = 3\times 2\cos 2t = 6\cos 2t$.

Similarly, the acceleration $a(t)$ is the rate of change of velocity at time t. That is

$$a(t) = \frac{dv(t)}{dt}.$$

So, in our example, $a(t) = \frac{dv(t)}{dt} = \frac{d6\cos 2t}{dt} = -12\sin 2t$.

Exercises and answers

Exercise 1

The coordinate x of a moving point at time t is given as

$$x(t) = t^4 - 3t^2 + 2t - 6.$$

Find the velocity and acceleration at time $t = 2$.

Exercise 2

A particle moves in such a way that at time t seconds its position x is given by

$$x(t) = 36 - 45t + 12t^2 - t^3.$$

Show that for the first 3 s its velocity is negative and then for the next 2 s its velocity is positive.

Exercise 3

A body is projected through a resistive medium and travels a distance x meters in a time t seconds according to the equation

$$x(t) = 36t - \frac{3}{8}t^2.$$

(1) How far does it travel before it comes to rest and how long does this take?

(2) What are the velocity $v(t)$ and acceleration $a(t)$ at time $t = 16$ s?

Answer 1

Since the distance traveled is given as $x(t) = t^4 - 3t^2 + 2t - 6$ so

$$v(t) = \frac{dx(t)}{dt} = 4t^3 - 6t + 2$$

and

$$a(t) = \frac{dv(t)}{dt} = 12t^2 - 6.$$

So that at $t = 2$

$$v(2) = 4 \times 2^3 - 6 \times 2 + 2 = 22$$

and

$$a(2) = 12 \times 2^2 - 6 = 42$$

Answer 2

Since $x(t) = 36 - 45t + 12t^2 - t^3$ so the velocity $v(t)$ is

$$v(t) = \frac{dx(t)}{dt} = -45 + 24t - 3t^2.$$

Now

$$v(t) = -45 + 24t - 3t^2 = -3(t^2 - 8t + 15) = -3(t-3)(t-5).$$

So, if $0 < t < 3$ $v(t)$ is negative, and for $3 < t < 5$ $v(t)$ is positive.

Answer 3

Given the distance as $x(t) = 36t - \frac{3}{8}t^2$ then

$$v(t) = 36 - \frac{6}{8}t$$

and

$$a(t) = -\frac{6}{8}.$$

(1) When the body comes to rest then $v(t) = 36 - \frac{6}{8}t = 0$. That is, when $t = 48$ s and when the body has traveled $x(48) = 36 \times 48 - \frac{3}{8} \times 48^2 = 864$ m.

(2) At time $t = 16$, $v(16) = 36 - \frac{6}{8} \times 16 = 24$ m s^{-1} and acceleration $a(t) = -\frac{3}{4}$ m s^{-1}s^{-1}.

Velocity and acceleration

You are now familiar with the fact that velocity is the rate of change of distance so that, for example, if the distance traveled by a moving particle $x(t)$ is given as an expression involving the time t as $x(t) = t^3$ then the velocity $v(t)$, being the rate of change of distance with respect to time is given as

$$v(t) = \frac{dx(t)}{dt} = \frac{dt^3}{dt} = 3t^2.$$

Furthermore, the acceleration $a(t)$, being the rate of change of velocity with respect to time, is given as

$$a(t) = \frac{dv(t)}{dt} = \frac{d3t^2}{dt} = 6t.$$

Now, can we turn this problem on its head? Given the acceleration can we deduce the velocity and hence the distance traveled? The answer is yes but we do need extra information as we shall see in the following example.

The acceleration $a(t)$ of a particle is given as an expression involving time t as $a(t) = 2t^3$. To find the velocity $v(t)$ we must reverse the process of differentiation and integrate. That is

$$v(t) = \int a(t)dt,$$

so for our example

$$v(t) = \int a(t)dt = \int 2t^3 dt = \frac{t^4}{2} + C.$$

Therefore, without any further information we can only find the velocity up to the addition of an arbitrary constant. To find the value of C we need to know the value of $v(t)$ for some particular value of t. If we find that at $t = 2$ then $v(2) = 6$ and we find that since

$$v(t) = \frac{t^4}{2} + C$$

then

$$v(2) = \frac{2^4}{2} + C = 6,$$

that is $v(2) = 8 + C = 6$ and so $C = -2$.

This gives the final answer as

$$v(t) = \frac{t^4}{2} - 2$$

and so we have fully determined the velocity at time t. We can go further and find the distance traveled by integrating the velocity. So that

$$x(t) = \int v(t)dt = \int\left(\frac{t^4}{2} - 2\right)dt = \frac{t^5}{10} - 2t + C.$$

Again we need further information before we can fully determine the distance traveled. Let us assume that $x(0) = 0$, i.e. at time $t = 0$ the distance traveled was also zero. In that case

$$x(0) = \frac{0^5}{10} - 2 \times 0 + C = C = 0$$

and so the distance is completely determined as

$$x(t) = \frac{t^5}{10} - 2t.$$

Exercises and answers

Exercise 1

A body travels with an acceleration given by $a(t) = 16 - 4t$ m s^{-2} with an initial velocity of 12 m s^{-1}. Obtain

(1) Its velocity $v(t)$
(2) The distance traveled $x(t)$ assuming $x(0) = 0$
(3) The distance traveled between $t = 1$ and $t = 4$

Exercise 2

Find the distance traveled, $x(t)$ m, and velocity, $v(t)$ m s^{-1}, at time t s if the acceleration is given as

(1) $a(t) = 3t$ and where $x(0) = 0$ m and $v(0) = 5$ m s^{-1}
(2) $a(t) = 12 - 6t$ and where $x(1) = 3$ m and $v(2) = 2$ m s^{-1}

Exercise 3

A particle, starting from rest, leaves at a point 5 m from point X and moves in a straight line away from X with a velocity of $v(t) = 64t - t^2$ m s^{-1}. Find

(1) The acceleration $a(t)$ and distance traveled from X.
(2) The time at which the particle begins to move back towards X and the time taken to return to X.

Exercise 4

The acceleration of a vehicle t s after starting from rest is given by $a(t) = 40 + 6t - t^2$ m s^{-2}. After 10 s the vehicle continues with a constant velocity. Find

(1) The time taken to attain the maximum speed
(2) The maximum speed.

Answer 1

(1) $v(t) = \int a(t)dt = \int (16 - 4t)dt = 16t - 2t^2 + C$. Now, $v(0) = C = 12$ and so $v(t) = 12 + 16t - 2t^2$.

(2) $x(t) = \int v(t)dt = \int (12 + 16t - 2t^2)dt = 12t + 8t^2 - \frac{2}{3}t^3 + C$. Now $x(0) = 0$ and so $C = 0$ giving

$$x(t) = 12t + 8t^2 - \frac{2}{3}t^3.$$

(3) $x(1) = 12 + 8 - \frac{2}{3} = 19\frac{1}{3}$ m and $x(4) = 48 + 128 - \frac{128}{3} = 133\frac{1}{3}$ m and so the distance traveled between $t = 1$ and $t = 4$ is 114 m.

Answer 2

(1) $a(t) = 3t$ so integrating $v(t) = \frac{3}{2}t^2 + C$. Now, $v(0) = 5$ and so $v(0) = C = 5$ therefore $v(t) = \frac{3}{2}t^2 + 5$ m s^{-1}. Integrating again, $x(t) = \frac{1}{2}t^3 + 5t + C$. Now, $x(0) = 0$ and so $C = 0$ giving $x(t) = \frac{1}{2}t^3 + 5t$ m.

(2) $a(t) = 12 - 6t$ so integrating $v(t) = 12t - 3t^2 + C$. Now, $v(2) = 2$ and so $v(2) = 24 - 12 + C = 2$, therefore $C = -10$ and so $v(t) = 12t - 3t^2 - 10$ m s^{-1}. Integrating again, $x(t) = 6t^2 - t^3 - 10t + C$. Now, $x(1) = 3$ and so $C = 8$ giving $x(t) = 6t^2 - t^3 - 10t + 8$ m.

Answer 3

(1) $v(t) = 64t - t^2$. Integration yields the distance traveled and differentiation yields the acceleration. Therefore $a(t) = 64 - 2t$ and $x(t) = 32t^2 - \frac{t^3}{3} + C$. Since we are considering the distance traveled from X we can say that $C = 0$. And so $x(t) = 32t^2 - \frac{t^3}{3}$.

(b) When $v(t) = 64t - t^2 = 0$ then the particle will have stopped. Since $64t - t^2 = t(64 - t)$ this occurs at $t = 0$ s and $t = 64$ s. At this time the acceleration is negative and the particle starts to move in the opposite direction, again with the acceleration $a(t) = 64 - 2t$. Since $x(t) = t^2(32 - \frac{t}{3})$ the particle returns to x when $32 - \frac{t}{3} = 0$, i.e. $t = 96$ s.

Answer 4

(1) The maximum speed is attained when the acceleration becomes zero and since $a(t) = 40 + 6t - t^2 = (10 - t)(4 + t)$ this occurs at $t = 10$ s.

(2) Given $a(t) = 40 + 6t - t^2$ then the speed can be found by integrating to give

$$v(t) = 40t + 3t^2 - \frac{t^3}{3}.$$

The integration constant is zero because the particle starts from rest. The maximum speed occurs $t = 10$ s so that

$$v(10) = 400 + 300 - \frac{1000}{3} = 366\frac{2}{3}\ \text{m s}^{-1}.$$

G3 METHODS OF INTEGRATION

Key Notes

Change of variable

On many occasions the expression to be integrated is not sufficiently simple to be in the tables of standard forms. It is therefore necessary to manipulate the integral so that it can be read from the table. This manipulation involves changing the variable.

Partial fractions

If the expression to be integrated is an algebraic fraction where the denominator can be factorized then partial fractions may be used to find a solution to the equation.

Change of variable

Many times we find that the expression that we wish to integrate is not sufficiently simple to be contained within our table of standard forms. For example, we may have to find the value of the integral

$$\int e^{3x}dx.$$

In our tables we can find the value of $\int e^{x}dx$ but here the x in the exponent is replaced with $3x$ so the table does not help at this stage. What we have to do is to manipulate the integral to get it into a form that can be read from the table. In this example, we let

$$u = 3x \text{ so that } du = \frac{du}{dx}dx = 3dx$$

and substitute this information into the integral to obtain

$$\int e^{3x}dx = \int e^{u}\frac{du}{3} = \frac{1}{3}\int e^{u}du.$$

This integral can now be evaluated using the table of standard forms:

$$\int e^{3x}dx = \frac{1}{3}\int e^{u}du = \frac{1}{3}\left(e^{u} + C_1\right) = \frac{1}{3}\left(e^{3x} + C_1\right) = \frac{e^{3x}}{3} + C,$$

where $C = {}^{C_1}\!/_3$. This is an example of a method called **change of variable** and we shall evaluate a few more to give you the idea of how it works.

Exercises and answers

Exercises 1 to 8

Evaluate each of the following integrals:

(1) $\int \sin 5x dx$ (2) $\int \ln(6x + 3)dx$

(3) $\int e^{2(x-4)}dx$ (4) $\int (x-4)^{1/2}dx$

(5) $\int x\cos(4+2x^2)dx$ (6) $\int 2x\ln(6x^2+3)dx$

(7) $\int \frac{dx}{\sqrt{x+1}}$ (8) $\int \frac{xdx}{\sqrt{x^2-3}}$

Answer 1

If $I = \int \sin 5xdx$: Let $u = 5x$ then $du = 5dx$ so $dx = du/5$, then I becomes

$$I = \int \sin u \frac{du}{5} = \frac{1}{5}\int \sin udu.$$

Using the table of standard forms we find that

$$I = -\frac{1}{5}\cos u + C = -\frac{1}{5}\cos 5x + C.$$

Answer 2

If $I = \int \ln(6x+3)dx$: Let $u = 6x+3$ then $du = 6dx$ so $dx = du/6$, then I becomes

$$I = \int \ln u \frac{du}{6} = \frac{1}{6}\int \ln udu.$$

Using the table of standard forms we find that

$$I = \frac{1}{6}(u \ln u - u) + C = \frac{1}{6}[(6x+3)\ln(6x+3) - (6x+3)] + C.$$

Answer 3

If $I = \int e^{2(x-4)}dx$: Let $u = 2(x-4)$ then $du = 2dx$ so $dx = du/2$, then I becomes

$$I = \int e^u \frac{du}{2} = \frac{1}{2}\int e^u du.$$

Using the table of standard forms we find that

$$I = \frac{e^u}{2} + C = \frac{e^{2(x-4)}}{2} + C.$$

Answer 4

If $I = \int (x-4)^{½}dx$: Let $u = (x-4)^{½}$ then $du = \frac{1}{2}(x-4)^{-½}dx = \frac{dx}{2u}$ so $dx = 2udu$, then I becomes

$$I = \int u2udu = 2\int u^2 du.$$

Using the table of standard forms we find that

$$I = \frac{2}{3}u^3 + C = \frac{2(x-4)^{3/2}}{3} + C.$$

Answer 5

If $I = \int x\cos(4+2x^2)dx$: Let $u = 4+2x^2$ then $du = 4xdx$ so $xdx = du/4$, then I becomes

$$I = \int \cos udu/4 = \frac{1}{4}\int \cos udu.$$

Using the table of standard forms we find that

$$I = \frac{1}{4}\sin u + C = \frac{\sin\left(4 + 2x^2\right)}{4} + C.$$

Answer 6

If $I = \int 2x \ln\left(6x^2 + 3\right)dx$: Let $u = 6x^2 + 3$ then $du = 12xdx$ so $2xdx = du/6$, then I becomes

$$I = \int \ln u du/6 = \frac{1}{6}\int \ln u du.$$

Using the table of standard forms we find that

$$I = \frac{1}{6}(u \ln u - u) + C = \frac{1}{6}\left[(6x^2 + 3) \ln (6x^2 + 3) - (6x^2 + 3)\right] + C.$$

Answer 7

If $I = \int \frac{dx}{\sqrt{x+1}}$: Let $u = \sqrt{x+1}$ then $du = \frac{1}{2\sqrt{x+1}}dx$ so $dx = 2udu$, then I becomes

$$I = \int \frac{2udu}{u} = 2\int du.$$

Using the table of standard forms we find that

$$I = 2u + C = 2\sqrt{x+1} + C.$$

Answer 8

If $I = \int \frac{xdx}{\sqrt{x^2-3}}$: let $u = \sqrt{x^2 - 3}$ then $du = \frac{x}{\sqrt{x^2-3}}dx$, so $xdx = udu$, then I becomes

$$I = \int \frac{udu}{u} = \int du.$$

Using the table of standard forms we find that

$$I = u + C = \sqrt{x^2 - 3} + C.$$

Partial fractions

If the expression to be integrated is an algebraic fraction where the denominator can be factorized then a method of solution using partial fractions may be appropriate. As a simple example consider the integral

$$\int \frac{dx}{x^2 + 5x + 6}$$

and let us pay particular attention to the expression being integrated, namely

$$\frac{1}{x^2 + 5x + 6} = \frac{1}{(x+2)(x+3)}.$$

The denominator has been factorized and we now make the assumption that the fraction can be alternatively written as a sum of partial fractions in the form

$$\frac{1}{(x+2)(x+3)} = \frac{A}{x+2} + \frac{B}{x+3},$$

where the values of A and B are to be found. This is done by adding the two partial fractions together again to obtain

$$\frac{1}{(x+2)(x+3)} = \frac{A}{x+2} + \frac{B}{x+3}$$
$$= \frac{A(x+3)+B(x+2)}{(x+2)(x+3)}.$$

The fraction on the left is equivalent to the fraction on the right and the denominators are identical. Therefore, their numerators must also be equivalent to each other. That is, $1 = A(x+3)+B(x+2)$. Furthermore, this equality must hold for any value of x. If we let $x = -2$ we find that

$$1 = A(-2+3)+B(-2+2) = A + B\times 0 = A.$$

If we now let $x = -3$ we find that

$$1 = A(-3+3)+B(-3+2) = A\times 0 + B(-1) = -B.$$

So we have found that $A = 1$ and $B = -1$ therefore

$$\frac{1}{(x+2)(x+3)} = \frac{1}{x+2} - \frac{1}{x+3}.$$

If this partial fraction breakdown is substituted into the integral we find that

$$\int \frac{dx}{x^2+5x+6} = \int \frac{dx}{(x+2)(x+3)}$$
$$= \int \left(\frac{1}{x+2} - \frac{1}{x+3} \right) dx$$
$$= \int \frac{dx}{x+2} - \int \frac{dx}{x+3}$$
$$= \ln(x+2) - \ln(x+3) + C \text{ (using the method of change of variable)}$$
$$= \ln\left(\frac{x+2}{x+3}\right) + C.$$

Exercises and answers

Exercises 1 to 3
Evaluate each of the following integrals:

(1) $\int \frac{dx}{x^2+2x-3}$

(2) $\int \frac{dx}{x^2-4}$

(3) $\int \frac{xdx}{8-10x-3x^2}$

Answer 1
Since $x^2+2x-3 = (x-1)(x+3)$ we find that

$$\frac{1}{x^2+2x-3} = \frac{1}{(x-1)(x+3)}$$
$$= \frac{A}{(x-1)} + \frac{B}{(x+3)}$$
$$= \frac{A(x+3)+B(x-1)}{(x-1)(x+3)}.$$

Now, by equating numerators we find that $1 = A(x+3)+B(x-1)$. Let $x = 1$ and we find that $1 = 4A + 0B$ and so $A = 1/4$.

Now let $x = -3$ and we find that $1 = 0A - 4B$ and so $B = -1/4$. Therefore,

$$\frac{1}{x^2+2x-3} = \frac{1/4}{(x-1)} - \frac{1/4}{(x+3)}.$$

Substituting into the integral gives

$$\int \frac{dx}{x^2+2x-3} = \frac{1}{4}\int \frac{dx}{x-1} - \frac{1}{4}\int \frac{dx}{x+3} = \frac{1}{4}(\ln(x-1) - \ln(x+3)) + C.$$

Answer 2

Since $x^2 - 4 = (x+2)(x-2)$ we find that

$$\begin{aligned} \frac{1}{x^2-4} &= \frac{1}{(x+2)(x-2)} \\ &= \frac{A}{(x+2)} + \frac{B}{(x-2)} \\ &= \frac{A(x-2)+B(x+2)}{(x+2)(x-2)}. \end{aligned}$$

Now, by equating numerators we find that $1 = A(x-2) + B(x+2)$. Let $x = -2$ and we find that $1 = -4A + 0B$ and so $A = -1/4$.

Now let $x = 2$ and we find that $1 = 0A + 4B$ and so $B = 1/4$. Therefore,

$$\frac{1}{x^2-4} = \frac{1/4}{(x-2)} - \frac{1/4}{(x+2)}.$$

Substituting into the integral gives

$$\int \frac{dx}{x^2-4} = \frac{1}{4}\int \frac{dx}{x-2} - \frac{1}{4}\int \frac{dx}{x+2} = \frac{1}{4}(\ln(x-2) - \ln(x+2)) + C.$$

Answer 3

Since $8 - 10x - 3x^2 = (2-3x)(4+x)$ we find that

$$\begin{aligned} \frac{x}{8-10x-3x^2} &= \frac{1}{(2-3x)(4+x)} \\ &= \frac{A}{(2-3x)} + \frac{B}{(4+x)} \\ &= \frac{A(4+x)+B(2-3x)}{(2-3x)(4+x)}. \end{aligned}$$

Now, by equating numerators we find that $x = A(4+x) + B(2-3x)$. Let $x = -4$ and we find that $-4 = 0A + 14B$ and so $B = -2/7$.

Now let $x = 2/3$ and we find that $2/3 = (14/3)A + 0B$ and so $A = 1/7$. Therefore,

$$\frac{x}{8-10x-3x^2} = \frac{1/7}{(2-3x)} - \frac{2/7}{(4+x)}.$$

Substituting into the integral gives

$$\int \frac{xdx}{8-10x-3x^2} = \frac{1}{7}\int \frac{dx}{(2-3x)} - \frac{2}{7}\int \frac{dx}{(4+x)} = -\frac{1}{21}\ln(2-3x) - \frac{2}{7}\ln(4+x) + C.$$

Division first

One overriding condition necessary for the partial fraction breakdown to work is that the degree of the numerator polynomial be greater than the degree of the denominator polynomial. For example, in the integral

$$\int \frac{x^2}{x^2 + 5x + 6} dx$$

we cannot immediate apply the previous method of partial fraction breakdown to the algebraic fraction

$$\frac{x^2}{x^2 + 5x + 6}$$

because the degree of the numerator is the same as the degree of the denominator, namely 2. In this case we must first divide as follows:

$$\begin{array}{r} 1 \\ x^2 + 5x + 6 \overline{) x^2 + 0x + 0} \\ \underline{x^2 + 5x + 6} \\ -5x - 6 \end{array}$$

Therefore

$$\frac{x^2}{x^2 + 5x + 6} = 1 - \frac{5x + 6}{x^2 + 5x + 6}$$

and the algebraic fraction on the right has a numerator of degree 1 less than the degree of the numerator so can be broken into partial fractions. Let's do this:

$$\begin{aligned} \frac{5x + 6}{x^2 + 5x + 6} &= \frac{5x + 6}{(x + 2)(x + 3)} \\ &= \frac{A}{x + 2} + \frac{B}{x + 3} \\ &= \frac{A(x + 3) + B(x + 2)}{(x + 2)(x + 3)}. \end{aligned}$$

The fraction on the left is equivalent to the fraction on the right and the denominators are identical. Therefore, their numerators must also be equivalent to each other. That is $5x + 6 = A(x + 3) + B(x + 2)$. Furthermore, this equality must hold for any value of x. If we let $x = -2$ we find that

$$-10 + 6 = -4 = A(-2 + 3) + B(-2 + 2) = A + B \times 0 = A.$$

If we now let $x = -3$ we find that

$$-15 + 6 = -9 = A(-3 + 3) + B(-3 + 2) = A \times 0 + B(-1) = -B.$$

So we have found that $A = -4$ and $B = 9$, therefore

$$\frac{5x + 6}{(x + 2)(x + 3)} = -\frac{4}{x + 2} + \frac{9}{x + 3}.$$

If this partial fraction breakdown is substituted into the integral we find that

$$\begin{aligned} \int \frac{x^2}{x^2 + 5x + 6} dx &= \int \left[1 - \left(-\frac{4}{x + 2} + \frac{9}{x + 3} \right) \right] dx \\ &= \int dx + 4\int \frac{dx}{x + 2} - 9\int \frac{dx}{x + 3} \\ &= x + 4\ln(x + 2) - 9\ln(x + 3) + C. \end{aligned}$$

Exercises and answers

Exercises 1 to 3

Evaluate each of the following integrals:

(1) $\int \frac{x^2}{x^2 - 3x + 2} dx$

(2) $\int \frac{x^4 - x^3}{x^2 + x - 20} dx$

(3) $\int \frac{x^2 + x + 20}{x^2 + x - 20} dx$

Answer 1

Because the degree of the numerator is the same as the degree of the denominator, namely 2, we must first divide as follows:

$$\begin{array}{r} 1 \\ x^2 - 3x + 2 \overline{)x^2 + 0x + 0} \\ \underline{x^2 - 3x + 2} \\ 3x - 2 \end{array}$$

Therefore:

$$\frac{x^2}{x^2 - 3x + 2} = 1 + \frac{3x - 2}{x^2 - 3x + 2}$$

and the algebraic fraction on the right has a numerator of degree 1 less than the degree of the numerator so can be broken into partial fractions. Let's do this:

$$\begin{aligned} \frac{3x - 2}{x^2 - 3x + 2} &= \frac{3x - 2}{(x - 1)(x - 2)} \\ &= \frac{A}{x - 1} + \frac{B}{x - 2} \\ &= \frac{A(x - 2) + B(x - 1)}{(x - 1)(x - 2)} \end{aligned}$$

The fraction on the left is equivalent to the fraction on the right and the denominators are identical. Therefore, their numerators must also be equivalent to each other. That is

$$3x - 2 = A(x - 2) + B(x - 1).$$

Furthermore, this equality must hold for any value of x. If we let $x = 1$ we find that

$$3 - 2 = 1 = A(1 - 2) + B(1 - 1) = -A + B \times 0 = -A.$$

If we now let $x = 2$ we find that

$$6 - 2 = 4 = A(2 - 2) + B(2 - 1) = A \times 0 + B(1) = B.$$

So we have found that $A = -1$ and $B = 4$, therefore

$$\frac{3x - 2}{(x - 1)(x - 2)} = -\frac{1}{x - 1} + \frac{4}{x - 2}.$$

If this partial fraction breakdown is substituted into the integral we find that

$$\begin{aligned}\int \frac{x^2}{x^2-3x+2}dx &= \int\left[1+\left(-\frac{1}{x-1}+\frac{4}{x-2}\right)\right]dx\\ &= \int dx + 4\int\frac{dx}{x-2} - \int\frac{dx}{x-1}\\ &= x + 4\ln(x-2) - \ln(x-1) + C.\end{aligned}$$

Answer 2

Because the degree of the numerator is the greater than the degree of the denominator we must first divide as follows:

$$\begin{array}{r l} & x^2 - 2x + 22 \\ x^2 + x - 20 \,\big) & x^4 - x^3 + 0x^2 + 0x + 0 \\ & \underline{x^4 + x^3 - 20x^2} \\ & -2x^3 + 20x^2 \\ & \underline{-2x^3 - 2x^2 + 40x} \\ & 22x^2 - 40x \\ & \underline{22x^2 + 22x - 440} \\ & -62x + 440 \end{array}$$

Therefore

$$\frac{x^4 - x^3}{x^2 + x - 20} = x^2 - 2x + 22 - \frac{62x - 440}{x^2 + x - 20}$$

and the algebraic fraction on the right has a numerator of degree 1 less than the degree of the numerator so can be broken into partial fractions. Let's do this:

$$\begin{aligned}\frac{62x-440}{x^2+x-20} &= \frac{62x-440}{(x-4)(x+5)}\\ &= \frac{A}{x-4}+\frac{B}{x+5}\\ &= \frac{A(x+5)+B(x-4)}{(x-4)(x+5)}\end{aligned}$$

The fraction on the left is equivalent to the fraction on the right and the denominators are identical. Therefore, their numerators must also be equivalent to each other. That is

$$62x - 440 = A(x+5) + B(x-4).$$

Furthermore, this equality must hold for any value of x. If we let $x = 4$ we find that

$$248 - 440 = -192 = A(4+5) + B(4-4) = 9A + B \times 0 = 9A.$$

If we now let $x = -5$ we find that

$$-310 - 440 = -750 = A(-5+5) + B(-5-4) = A \times 0 + B(-9) = -9B.$$

So we have found that $A = -\frac{192}{9}$ and $B = \frac{750}{9}$, therefore

$$\frac{62x-440}{x^2+x-20} = \frac{1}{9}\left(\frac{750}{x+5} - \frac{192}{x-4}\right).$$

If this partial fraction breakdown is substituted into the integral we find that

$$\begin{aligned}\int \frac{x^4 - x^3}{x^2 + x - 20}\,dx &= \int\left[x^2 - 2x + 22 - \frac{1}{9}\left(\frac{750}{x+5} - \frac{192}{x-4}\right)\right]dx \\ &= \int (x^2 - 2x + 22)dx - \frac{750}{9}\int \frac{dx}{x+5} + \frac{192}{9}\int \frac{dx}{x-4} \\ &= \frac{x^3}{3} - x^2 + 22x - \frac{750}{9}\ln(x+5) + \frac{192}{9}\ln(x-4) + C.\end{aligned}$$

Answer 3
Since $x^2 + x + 20 = x^2 + x - 20 + 40$,

$$\frac{x^2 + x + 20}{x^2 + x - 20} = \frac{x^2 + x - 20 + 40}{x^2 + x - 20} = 1 + \frac{40}{x^2 + x - 20}.$$

Then

$$\begin{aligned}\frac{40}{x^2 + x - 20} &= \frac{40}{(x-4)(x+5)} \\ &= \frac{A}{x-4} + \frac{B}{x+5} \\ &= \frac{A(x+5) + B(x-4)}{(x-4)(x+5)}.\end{aligned}$$

The fraction on the left is equivalent to the fraction on the right and the denominators are identical. Therefore, their numerators must also be equivalent to each other. That is $40 = A(x+5) + B(x-4)$. Furthermore, this equality must hold for any value of x. If we let $x = 4$ we find that

$$40 = A(4+5) + B(4-4) = 9A + B\times 0 = 9A.$$

If we now let $x = -5$ we find that

$$40 = A(-5+5) + B(-5-4) = A\times 0 + B(-9) = -9B.$$

So we have found that $A = \frac{40}{9}$ and $B = -\frac{40}{9}$ and therefore

$$\frac{40}{x^2 + x - 20} = \frac{40}{9}\left(\frac{1}{x-4} - \frac{1}{x+5}\right)$$

so that

$$\frac{x^2 + x + 20}{x^2 + x - 20}\,dx = x + \frac{40}{9}\left[\ln(x-4) - \ln(x+5)\right] + C.$$

Repeated factors
When repeated factors occur in the denominator these have to be catered for in the subsequent partial fraction breakdown. As an example consider the algebraic fraction

$$\frac{9}{(x+1)(x-2)^2}.$$

Here the repeated factor $(x-2)$ is catered for by assuming a partial fraction breakdown of the form

$$\frac{9}{(x+1)(x-2)^2} = \frac{A}{x+1} + \frac{B}{x-2} + \frac{C}{(x-2)^2}.$$

Adding the fractions together gives

$$= \frac{A(x-2)^2 + B(x+1)(x-2) + C(x+1)}{(x+1)(x-2)^2}.$$

Therefore

$$9 = A(x-2)^2 + B(x+1)(x-2) + C(x+1).$$

Letting $x = -1$ yields $A = 1$, and letting $x = 2$ yields $C = 3$. If now $x = 0$ we find $9 = 4 - 2B + 3$ and so $B = -1$, yielding the partial fraction breakdown as

$$\frac{9}{(x+1)(x-2)^2} = \frac{1}{x+1} - \frac{1}{x-2} + \frac{3}{(x-2)^2}.$$

Exercises and answers

Exercises 1 to 3
Evaluate each of the following integrals:

(1) $\int \frac{9}{(x+1)(x-2)^2} dx$

(2) $\int \frac{dx}{(x^2-1)(x+1)}$

(3) $\int \frac{x^2}{(2x+1)(x+3)^2} dx$

Answer 1
We have already seen that

$$\frac{9}{(x+1)(x-2)^2} = \frac{1}{x+1} - \frac{1}{x-2} + \frac{3}{(x-2)^2}$$

and so

$$\begin{aligned}\int \frac{9}{(x+1)(x-2)^2} dx &= \int \left(\frac{1}{x+1} - \frac{1}{x-2} + \frac{3}{(x-2)^2} \right) dx \\ &= \int \frac{dx}{x+1} - \int \frac{dx}{x-2} dx + 3\int \frac{dx}{(x-2)^2} \\ &= \ln(x+1) - \ln(x-2) - 3(x-2)^{-1} + C.\end{aligned}$$

Answer 2
We know that

$$\begin{aligned}\frac{1}{(x^2-1)(x+1)} &= \frac{1}{(x-1)(x+1)(x+1)} \\ &= \frac{1}{(x-1)(x+1)^2} \\ &= \frac{A}{x-1} + \frac{B}{x+1} + \frac{C}{(x+1)^2} \\ &= \frac{A(x+1)^2 + B(x-1)(x+1) + C(x-1)}{(x-1)(x+1)^2}.\end{aligned}$$

and so $1 = A(x+1)^2 + B(x-1)(x+1) + C(x-1)$.

- Let $x = 1$ to give $1 = 4A$, yielding $A = 1/4$.
- Let $x = -1$ to give $1 = -2C$, yielding $C = -1/2$.

- Let $x = 0$ to give $1 = \frac{1}{4} - B + \frac{1}{2}$ yielding $B = -1/4$.

Therefore

$$\frac{1}{(x^2-1)(x+1)} = \frac{1/4}{x-1} - \frac{1/4}{x+1} - \frac{1/2}{(x+1)^2}$$

and so

$$\begin{aligned}\int \frac{dx}{(x^2-1)(x+1)} &= \int \left(\frac{1/4}{x-1} - \frac{1/4}{x+1} - \frac{1/2}{(x+1)^2} \right) dx \\ &= \frac{1}{4}\int \frac{dx}{x-1} - \frac{1}{4}\int \frac{dx}{x+1} - \frac{1}{2}\int \frac{dx}{(x+1)^2} \\ &= \frac{1}{4}\ln(x-1) - \frac{1}{4}\ln(x+1) + \frac{1}{2}(x+1)^{-1} + C.\end{aligned}$$

Answer 3
We know that

$$\begin{aligned}\frac{x^2}{(2x+1)(x+3)^2} &= \frac{A}{2x+1} + \frac{B}{x+3} + \frac{C}{(x+3)^2} \\ &= \frac{A(x+3)^2 + B(2x+1)(x+3) + C(2x+1)}{(2x+1)(x+3)^2}.\end{aligned}$$

Therefore

$$x^2 = A(x+3)^2 + B(2x+1)(x+3) + C(2x+1).$$

- Let $x = -3$ to give $9 = -5C$, yielding $C = -9/5$.
- Let $x = -1/2$ to give $\frac{1}{4} = \frac{25}{4}A$ yielding $A = 1/25$.
- Let $x = 0$ to give $0 = \frac{9}{25} + 3B - \frac{9}{5}$ yielding $B = -12/25$.

Hence

$$\begin{aligned}\int \frac{x^3}{(2x+1)(x+3)^2} dx &= \int \left(\frac{1/25}{2x+1} + \frac{12/25}{x+3} - \frac{9/5}{(x+3)^2} \right) dx \\ &= \frac{1}{25}\int \frac{dx}{2x+1} + \frac{12}{25}\int \frac{dx}{x+3} - \frac{9}{5}\int \frac{dx}{(x+3)^2} \\ &= \frac{1}{25}\left[\frac{1}{2}\ln(2x+1)\right] + \frac{12}{25}\ln(x+3) + \frac{9}{5}(x+3)^{-1} + C.\end{aligned}$$

G4 AREAS UNDER LINES

In *Fig. 1*, A is the area enclosed below the graph of $f(x)$, above the x-axis and between the two vertical lines $x = a$ and $x = b$. To find the area A we need first of all to find the total area beneath the same curve from the left up to some arbitrary point P on the curve where P has the coordinates (x, y). We shall denote this area by the symbol A_x (*Fig. 2*).

The area δA_x is the area enclosed in the strip beneath the arc PQ where Q has the coordinates $(x + \delta x, y + \delta y)$. If the area of the strip is approximated by a rectangle of height y and width δx then

$$\delta A_x \simeq y\delta x, \text{ so that } \frac{\delta A_x}{\delta x} \simeq y.$$

The error in this approximation is equal to the area of PQR and if the width of the strip is reduced the error is also reduced. Also, if δx approaches zero so δA_x also approaches zero but the ratio $\frac{\delta A_x}{\delta x}$ approaches both the value of y and the

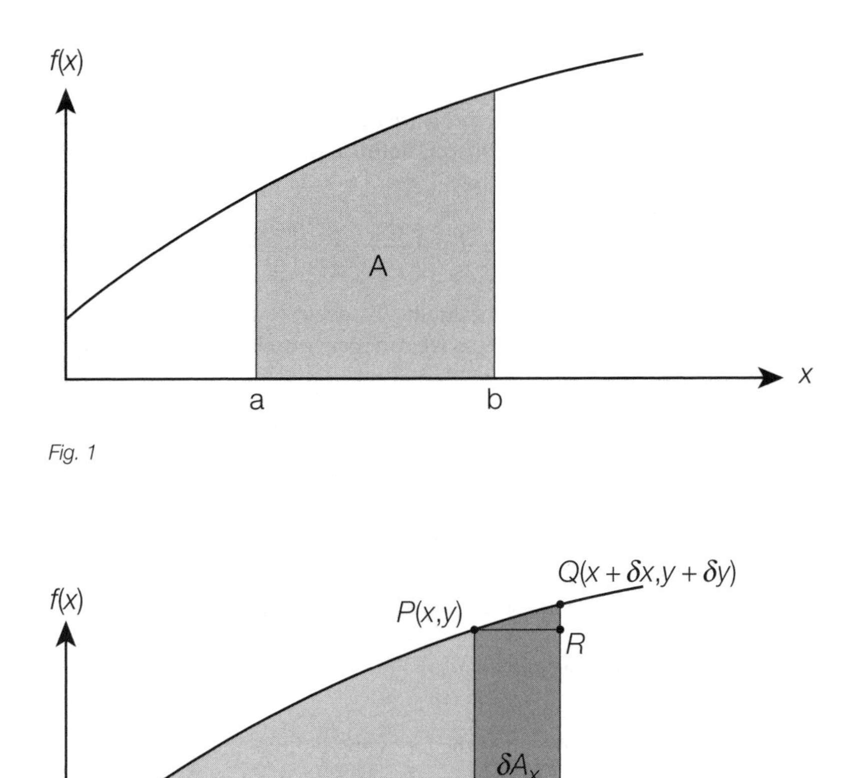

Fig. 1

Fig. 2

value of the derivative $\frac{dA_x}{dx}$. We conclude that $\frac{dA_x}{dx} = y$ and because integration is the reverse process of differentiation we can see that

$$A_x = \int y dx.$$

The total area between the curve and the x-axis up to the point P is given by the indefinite integral.

If $x = b$ then

$$A_b = \int_{(x=b)} y dx,$$

the area up to $x = b$; and if $x = a$ then

$$A_a = \int_{(x=a)} y dx,$$

the area up to $x = a$. Consequently, the area between the two vertical lines $x = a$ and $x = b$ is given as

$$A_b - A_a = \int_{(x=b)} y dx - \int_{(x=a)} y dx$$

and we write this as

$$A = \int_a^b y dx.$$

The numbers a and b are called the lower and upper limits of the integral. Such an integral is called a **definite integral** because it is evaluated between definite limits. For example,

$$\int_2^3 x^2 dx = \left[\frac{x^3}{3} + C\right]_2^3 = \left(\frac{3^3}{3} + C\right) - \left(\frac{2^3}{3} + C\right) = 9 - \frac{8}{3} = \frac{19}{3}.$$

Notice that the value if C is irrelevant because it is added then subtracted. This always happens so we can ignore it altogether and simply write

$$\int_2^3 x^2 dx = \left[\frac{x^3}{3}\right]_2^3 = \left(\frac{3^3}{3}\right) - \left(\frac{2^3}{3}\right) = 9 - \frac{8}{3} = \frac{19}{3}.$$

We shall do one more and then you can try some for yourself.

$$\int_0^1 \frac{dx}{x+1} = [\ln(x+1)]_0^1 = (\ln 2) - (\ln 1) = \ln 2$$

because $\ln 1 = 0$. Try the following exercises for yourself.

Exercises and answers

Exercises 1 to 3

Evaluate each of the following definite integrals:

(1) $\int_0^1 e^{3x} dx$

(2) $\int_{-2}^2 (4x + 2)^2 dx$

(3) $\int_2^4 \frac{dx}{2x-3}$

Answer 1

Integration gives

$$\int_0^1 e^{3x}dx = \left[\frac{e^{3x}}{3}\right]_0^1 = \left(\frac{e^3}{3} - \frac{e^0}{3}\right) = \frac{1}{3}(e^3 - 1)$$

Answer 2

Integration gives

$$\int_{-2}^{2} (4x+2)^2 dx = \left[\frac{(4x+2)^3}{12}\right]_{-2}^{2} = \left(\frac{1000}{12} - \frac{(-216)}{12}\right) = \frac{1216}{12} = \frac{304}{3}$$

Answer 3

Integration gives

$$\int_2^4 \frac{dx}{2x-3} = \left[\frac{\ln(2x-3)}{2}\right]_2^4 = \left(\frac{\ln 5}{2} - \frac{\ln 1}{2}\right) = \frac{1}{2}\ln 5 = \ln\sqrt{5}$$

G5 NUMERICAL INTEGRATION

Key Notes

Trapezoidal rule

A method of numerical integration developed by approximating the area beneath a curve as a series of trapezia. It is useful when we are faced with an integral whose solution will not succumb to any method.

Simpson's rule

An extension of the trapezoidal rule which involves joining the three end points of two adjacent strips with a parabola.

The trapezoidal rule

Very often we are faced with an integral whose solution will not succumb to any method. In cases such as these we have to resort to numerical methods and we shall discuss just two of these, the trapezoidal rule and Simpson's rule, here.

The **trapezoidal rule** for numerical integration is developed by approximating the area beneath a curve by a series of trapezia – four-sided figures where two of the sides are parallel. This is done by subdividing the x-axis into equal intervals a, x_1, x_2, ..., b as shown in *Fig. 1*.

Vertical lines are constructed from these points on the x-axis up to the curve and then capped off as shown to form a series of inscribed trapezia. The area of each trapezium is given as the average length of the two parallel sides multiplied by their separation. This separation is given as

$$w = \frac{b-a}{n}$$

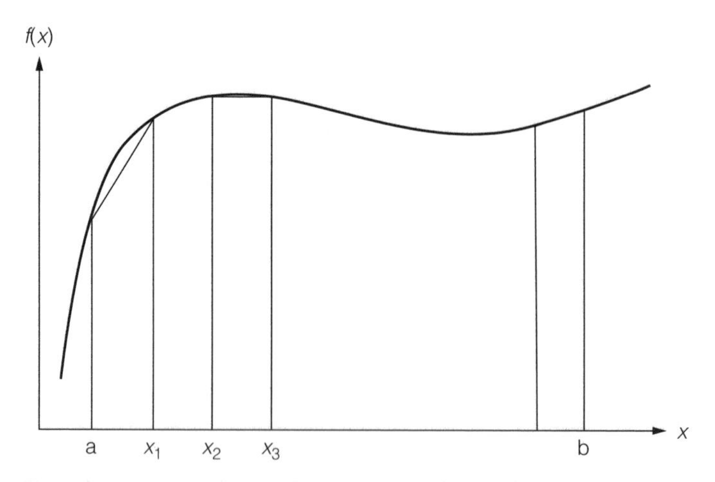

Fig. 1. Approximating the area beneath a curve by a series of trapezia.

where n is the number of strips. By adding up the areas of all the inscribed trapezia we arrive at the formula for the trapezoidal rule for approximate integration:

$$A = \frac{w}{2}\{f(a) + 2[f(x_1) + f(x_2) + \ldots] + f(b)\}$$

Here a and b are the end points of the integral, the points x_1, x_2, ... form the subdivision of the x-axis and w is the width of each interval. Let's do an example to see how this works. We shall use the trapezium rule to evaluate the integral

$$\int_0^{\pi/2} x \sin x dx$$

with four divisions of the x-axis, Firstly, we divide the interval of the x-axis $[0, \pi/2]$ into four equal subdivisions. These are $[0, \pi/8]$, $[\pi/8, \pi/4]$, $[\pi/4, 3\pi/8]$ and $[3\pi/8, \pi/2]$. The width w of each subdivision is $\pi/8$ and so, substituting this information into the formula we find that

$$\begin{aligned} A &= \frac{w}{2}\{f(a) + 2[f(x_1) + f(x_2) + \ldots] + f(b)\} \\ &= \frac{\pi/8}{2}\{0 \sin 0 + 2[(\pi/8)\sin(\pi/8) + (\pi/4)\sin(\pi/4) + \\ &\quad (3\pi/8)\sin(3\pi/8)] + (\pi/2)\sin(\pi/2)\} \\ &= \frac{0.3927}{2}\{0 + 2[0.1503 + 0.5554 + 1.0884] + 1.5708\} \\ &= 0.1964\{5.159\} \\ &= 1.013 \quad \text{to three decimal places.} \end{aligned}$$

Simpson's rule

Simpson's rule gives an improvement on the trapezoidal rule but its derivation is not as obvious because it involves successively joining the three end points of two adjacent strips with a parabola. We cite the formula

$$A = \frac{w}{3}\{f(a) + 4[f(x_1) + f(x_3) + \ldots] + 2[f(x_2) + f(x_4) + \ldots] + f(b)\}$$

For example, we shall use Simpson's rule to evaluate the integral

$$\int_0^{\pi/2} x \sin x dx$$

with four divisions of the x-axis.

Again we divide the interval of the x-axis $[0, \pi/2]$ into four equal subdivisions. These are $[0, \pi/8]$, $[\pi/8, \pi/4]$, $[\pi/4, 3\pi/8]$ and $[3\pi/8, \pi/2]$. The width w of each subdivision is $\pi/8$ and so, substituting this information into the formula we find that

$$\begin{aligned} A &= \frac{w}{3}\{f(a) + 4[f(x_1) + f(x_3) + \ldots] + 2[f(x_2) + f(x_4) + \ldots] + f(b)\} \\ &= \frac{\pi/8}{3}\{0 \sin 0 + 4[(\pi/8)\sin(\pi/8) + (3\pi/8)\sin(3\pi/8)] + \\ &\quad 2[(\pi/4)\sin(\pi/4)] + (\pi/2)\sin(\pi/2)\} \\ &= \frac{0.3927}{3}\{0 + 4[0.1503 + 1.0884] + 2[0.5554] + 1.5708\} \\ &= 0.1309\{7.6364\} \\ &= 0.9996 \text{ to four decimal places.} \end{aligned}$$

If we were to evaluate this integral exactly then we would find that the its value was 1 so that Simpson's rule provides a more accurate result.

Exercises and answers

Exercises 1 to 3

Evaluate each of the following integrals with six divisions of the x-axis, first by the trapezoidal rule and then by Simpson's rule and then compare with the exact value:

(1) $\int_0^{\pi} \sin x dx$

(2) $\int_{-1}^{1} e^x dx$

(3) $\int_2^4 \frac{dx}{x^2-1}$

Answer 1

Writing $f(x) = \sin x$ we find that the six divisions of the x-axis are, to three decimal places ($w = \pi/6$):

x	0	$\pi/6$	$\pi/3$	$\pi/2$	$2\pi/3$	$5\pi/6$	π
$f(x)$	0	0.500	0.866	1.000	0.866	0.500	0

By the trapezoidal rule

$$A = \frac{w}{2}\{f(a) + 2[f(x_1) + f(x_2) + \ldots] + f(b)\}$$

we find that

$$A = \frac{\pi}{12}\{0 + 2[0.5 + 0.866 + 1 + 0.866 + 0.5] + 0\}$$
$$= 1.954 \quad \text{to three decimal places.}$$

By Simpson's rule

$$A = \frac{w}{3}\{f(a) + 4[f(x_1) + f(x_3) + \ldots] + 2[f(x_2) + f(x_4) + \ldots] + f(b)\}$$

we find that

$$A = \frac{\pi}{18}\{0 + 4[0.5 + 1 + 0.5] + 2[0.866 + 0.866] + 0\}$$
$$= 2.001 \quad \text{to three decimal places.}$$

Exactly:

$$\int_0^{\pi} \sin x dx = [-\cos x]_0^{\pi} = ((-\cos\pi) - (-\cos 0)) = (1 - (-1)) = 2.$$

Answer 2

Writing $f(x) = e^x$ we find that the six divisions of the x-axis are, to three decimal places ($w = 2/6 = 1/3$):

x	−1	−2/3	−1/3	0	1/3	2/3	1
$f(x)$	0.368	0.513	0.717	1.000	1.396	1.948	2.718

By the trapezoidal rule:

$$A = \frac{w}{2}\{f(a) + 2[f(x_1) + f(x_2) + \ldots] + f(b)\}$$

we find that

$$A = \frac{1}{6}\{0.368 + 2[0.513 + 0.717 + 1 + 1.396 + 1.948] + 2.718\}$$
$$= 2.372 \quad \text{to three decimal places.}$$

By Simpson's rule:

$$A = \frac{w}{3}\{f(a) + 4[f(x_1) + f(x_3) + \ldots] + 2[f(x_2) + f(x_4) + \ldots] + f(b)\}$$

we find that

$$A = \frac{1}{9}\{0.368 + 4[0.513 + 1 + 1.948] + 2[0.717 + 1.396] + 2.718\}$$
$$= 2.351 \quad \text{to three decimal places.}$$

Exactly, by integrating:

$$\int_{-1}^{1} e^x dx = [e^x]_{-1}^{1} = (e - e^{-1}) = 2.718 - 0.368 = 2.350.$$

Answer 3

Writing $f(x) = \frac{1}{x^2 - 1}$ we find that the six divisions of the x-axis are, to three decimal places ($w = 2/6 = 1/3$):

x	2	$2\frac{1}{3}$	$2\frac{2}{3}$	3	$3\frac{1}{3}$	$3\frac{2}{3}$	4
$f(x)$	$\frac{1}{3}$	$\frac{9}{40}$	$\frac{9}{55}$	$\frac{1}{8}$	$\frac{9}{91}$	$\frac{9}{112}$	$\frac{1}{15}$

By the trapezoidal rule:

$$A = \frac{w}{2}\{f(a) + 2[f(x_1) + f(x_2) + \ldots] + f(b)\}$$

we find that

$$A = \frac{1}{6}\left\{\frac{1}{3} + 2\left[\frac{9}{40} + \frac{9}{55} + \frac{1}{8} + \frac{9}{91} + \frac{9}{112}\right] + \frac{1}{15}\right\},$$

that is:

$$A = 0.167\{0.333 + 2[0.225 + 0.164 + 0.125 + 0.099 + 0.080] + 0.067\}$$
$$= 0.298 \text{ to three decimal places.}$$

By Simpson's rule:

$$A = \frac{w}{3}\{f(a) + 4[f(x_1) + f(x_3) + \ldots] + 2[f(x_2) + f(x_4) + \ldots] + f(b)\}$$

we find that

$$A = \frac{1}{9}\{0.333 + 4[0.225 + 0.125 + 0.080] + 2[0.164 + 0.099] + 0.067\}$$
$$= 0.294 \quad \text{to three decimal places.}$$

Exactly, by integrating: by partial fractions we find that

$$\frac{1}{x^2-1}=\frac{1}{(x+1)(x-1)}=\frac{1/2}{x-1}-\frac{1/2}{x+1}$$

and so

$$\begin{aligned}\int_2^4\frac{dx}{x^2-1}&=\frac{1}{2}\int_2^4\frac{dx}{x-1}-\frac{1}{2}\int_2^4\frac{dx}{x+1}\\&=\frac{1}{2}[\ln(x-1)]_2^4-\frac{1}{2}[\ln(x+1)]_2^4\\&=\frac{1}{2}(\ln 3-\ln 1)-\frac{1}{2}(\ln 5-\ln 3)\\&=\ln 3-\ln\sqrt{5}=0.294\quad\text{to three decimal places.}\end{aligned}$$

H1 DIFFERENTIAL EQUATIONS

Key Notes

What is a differential equation?
A differential equation is any equation that contains a function and its derivatives. They can be first order or second order.

Solving differential equations by integration
If we start with a differential equation then we can use integration to find the function that satisfies the equation.

Separation of variables
A technique for solving differential equations when direct integration is not possible.

What is a differential equation?

A differential equation is any equation that contains a function and its derivatives. The following are typical examples:

$$\frac{df(x)}{dx} - 3x^2 f(x) = e^{-4x},$$

$$\frac{d^2g(t)}{dt^2} + \sin t \frac{dg(t)}{dt} - 4tg(t) = \tan t.$$

The first example is a **first-order differential equation** because the highest derivative it contains is the first. The second is a **second-order differential equation**. The purpose of solving a differential equation is to find the function that satisfies it. In the two above equations they would be solved when we had found the function $f(x)$ that satisfied the first one and the function $g(t)$ that satisfied the second one.

Differential equations can be used to describe our understanding of **physical behavior**. For example, let a particle move along the x-axis so that its distance from the point $x = 0$ is given by the equation $x(t) = \sin t$, where t represents time. Because

$$\frac{dx(t)}{dt} = \cos t$$

and

$$\frac{d^2x(t)}{dt^2} = \frac{d}{dt}\left(\frac{dx(t)}{dt}\right) = -\sin t = -x(t)$$

we can say that the motion of the particle satisfies the differential equation

$$\frac{d^2x(t)}{dt^2} = -x(t).$$

Exercises and answers

Exercises 1 to 3

Find the differential equations satisfied by each of the following functions:

(1) $f(x) = \cos x$
(2) $g(x) = e^{3x}$
(3) $h(x) = 6x$

Answer 1

Since $\frac{df(x)}{dx} = -\sin x$ and $\frac{d^2f(x)}{dx^2} = -\cos x = -f(x)$ the differential equation is

$$\frac{d^2f(x)}{dx^2} = -f(x) \text{ or } \frac{d^2f(x)}{dx^2} + f(x) = 0.$$

Answer 2

Since $\frac{dg(x)}{dx} = 3e^{3x} = 3g(x)$ the differential equation is

$$\frac{dg(x)}{dx} = 3g(x) \text{ or } \frac{dg(x)}{dx} - 3g(x) = 0.$$

Answer 3

Since $\frac{dh(x)}{dx} = 6 = \frac{h(x)}{x}$ the differential equation is

$$\frac{dh(x)}{dx} = \frac{h(x)}{x} \text{ or } \frac{dh(x)}{dx} - \frac{h(x)}{x} = 0.$$

Solving differential equations by integration

So far we have started with a function and then found the differential equation it satisfies. What we want to do now is to go the other way. We want to start with the differential equation and find the function that satisfies it: we wish to solve the differential equation and this is a much harder thing to do.

We know that integration is the reverse process of differentiation so the process of solving a differential equation – a process of undoing the differentiation – involves integration. The simplest differential equations are those that can be solved by direct integration. For example, to solve

$$\frac{df(x)}{dx} = 4x^2$$

we simply integrate to obtain

$$\int \frac{df(x)}{dx} dx = \int df(x) = f(x) = \int 4x^2 dx = \frac{4}{3}x^3 + C.$$

That is, $f(x) = \frac{4}{3}x^3 + C$. To find the values of the integration constant, C, we need to know the value of the function for some specific value of x. Such values are given separately from the differential equation and are variously known as **boundary values** or **initial conditions**.

For example, solve the differential equation

$$\frac{df(x)}{dx} = 6 \sin 2x$$

given that $f(0) = 3$. Again, we integrate directly to obtain $f(x) = \int 6\sin 2x dx = -3\cos 2x + C$. That is, $f(x) = -3\cos 2x + C$. Now, $f(0) = -3\cos 0 + C$. Since $\cos 0 = 1$ this means that $3 = -3 + C$, giving $C = 6$. The complete solution to the differential equation is then $f(x) = -3\cos 2x + 6$. So try a few yourself.

Exercises and answers

Exercises 1 to 5

By direct integration and, where appropriate, the application of the boundary condition solve each of the following differential equations:

(1) $\dfrac{df(x)}{dx} = 5x^3$

(2) $\dfrac{ds(t)}{dt} = 7t^4$ where $s(0) = 1$

(3) $\dfrac{dg(x)}{dx} = 2e^{3x}$

(4) $\dfrac{dh(t)}{dt} = 9e^{-5t}$ where $h(0) = 0$

(5) $\dfrac{df(x)}{dx} = 2x + 3e^{-x} - 4\sin 5x$ where $f(0) = 1$

Answer 1

As $\dfrac{df(x)}{dx} = 5x^3$ so $f(x) = \int 5x^3 dx = \dfrac{5}{4}x^4 + C$. That is, $f(x) = \dfrac{5}{4}x^4 + C$.

Answer 2

As $\dfrac{ds(t)}{dt} = 7t^4$ so $s(t) = \int 7t^4 dt = \dfrac{7}{5}t^5 + C$. That is, $s(t) = \dfrac{7}{5}t^5 + C$.

Now, $s(0) = 1$ and so so $s(0) = \dfrac{7}{5}0^5 + C = C = 1$. Therefore $s(t) = \dfrac{7}{5}t^5 + 1$.

Answer 3

As $\dfrac{dg(x)}{dx} = 2e^{3x}$ so $g(x) = \int 2e^{3x} dx = \dfrac{2}{3}e^{3x} + C$. That is, $g(x) = \dfrac{2}{3}e^{3x} + C$.

Answer 4

As $\dfrac{dh(t)}{dt} = 9e^{-5t}$ so $h(t) = \int 9e^{-5t} dt = -\dfrac{9}{5}e^{-5t} + C$. That is $h(t) = -\dfrac{9}{5}e^{-5t} + C$.

Now, $h(0) = 0$ and so $h(0) = -\dfrac{9}{5}e^0 + C = -\dfrac{9}{5} + C = 0$, since $e^0 = 1$. Therefore $C = \dfrac{9}{5}$, giving the final solution as:

$$h(t) = \frac{9}{5}\left(1 - e^{-5t}\right).$$

Answer 5

As $\dfrac{df(x)}{dx} = 2x + 3e^{-x} - 4\sin 5x$ so $f(x) = \int 2xdx + \int 3e^{-x}dx - \int 4\sin 5xdx$. That is, $f(x) = x^2 - 3e^{-x} + \dfrac{4}{5}\cos 5x + C$.

Now, $f(0) = 1$ and so $f(0) = 0 - 3 + \dfrac{4}{5} + C = 1$, giving $C = \dfrac{16}{5}$.

The final solution is then

$$f(x) = x^2 - 3e^{-x} + \frac{4}{5}\cos 5x + \frac{16}{5}.$$

Second-order differential equations

Certain **second-order differential equations** can be solved in a similar manner, remembering that the **second derivative is the derivative of the first derivative**. So that, for example, the solution of the differential equation

$$\frac{d^2f(x)}{dx^2} = 2x^3$$

can be found by integrating twice. That is,

$$\int \frac{d^2f(x)}{dx^2}dx = \int 2x^3dx$$

so that, since the integral of the second derivative is the first derivative:

$$\frac{df(x)}{dx} = \frac{x^4}{2} + C_1,$$

giving

$$f(x) = \int\left(\frac{x^4}{2} + C_1\right)dx = \frac{x^5}{10} + C_1x + C_2.$$

That is,

$$f(x) = \frac{x^5}{10} + C_1x + C_2.$$

To find the values of C_1 and C_2 we need two boundary conditions. Consider the following example. Solve the differential equation

$$\frac{d^2p(t)}{dt^2} = 7e^{-3t}$$

where $p(0) = 0$ and

$$\left.\frac{dp(t)}{dt}\right|_{t=0} = 1$$

(the derivative of $p(t)$ when $t = 0$ is 1).

Solution

As

$$\frac{d^2p(t)}{dt^2} = 7e^{-3t}$$

so

$$\frac{dp(t)}{dt} = -\frac{7}{3}e^{-3t} + C_1.$$

Now,

$$\left.\frac{dp(t)}{dt}\right|_{t=0} = 1$$

and so

$$1 = -\frac{7}{3} + C_1,$$

$$\text{making } C_1 = \frac{10}{3}.$$

Therefore

$$\frac{dp(t)}{dt} = -\frac{7}{3}e^{-3t} + \frac{10}{3}.$$

Hence, by integration

$$p(t) = \frac{7}{9}e^{-3t} + \frac{10}{3}t + C_2.$$

The last boundary condition is $p(0) = 0$ and so

$$p(0) = \frac{7}{9} + C_2 = 0,$$

giving

$$C_2 = -\frac{7}{9}$$

and a final solution of

$$\begin{aligned} p(t) &= \frac{7}{9}e^{-3t} + \frac{10}{3}t - \frac{7}{9} \\ &= \frac{7}{9}\left(e^{-3t} - 1\right) + \frac{10}{3}t. \end{aligned}$$

Solving differential equations by separating the variables

Sometimes it may not be possible to solve a differential equation by direct integration immediately. Instead, some manipulation of the equation may be necessary before it is possible to integrate. Such is the case where **separating the variables** is possible. Consider, as an example, the differential equation:

$$\frac{df(x)}{dx} = f(x).$$

Because we do not know $f(x)$ we cannot integrate this equation as it stands. Instead we separate terms involving $f(x)$ from the term involving just x as follows:

$$\frac{df(x)}{f(x)} = dx.$$

We now change the variable by letting $f(x) = y$ to give $\frac{dy}{y} = dx$. We now integrate both sides of this equation $\int \frac{dy}{y} = \int dx$. That is, $\ln y = x + C$. That is, $y = e^{x+C} = e^x e^C = Ae^x$, where $A = e^C$. Reverting to the original notation we see that $f(x) = Ae^x$ is the solution.

Try another. Solve the equation

$$\frac{dy}{dx} = \frac{x}{y}$$

given that when $x = 2$, $y = -3$.

Solution

Given $\frac{dy}{dx} = \frac{x}{y}$ we separate the variables to find $ydy = xdx$. Integrating this equation gives $\int ydy = \int xdx$; that is,

$$\frac{y^2}{2} = \frac{x^2}{2} + C.$$

Applying the boundary condition $x = 2$, $y = -3$ we see that

$$\frac{9}{2} = \frac{4}{2} + C$$

and so

$$C = \frac{5}{2}.$$

Therefore

$$\frac{y^2}{2} = \frac{x^2}{2} + \frac{5}{2},$$

so $y^2 = x^2 + 5$, giving $y = \pm\sqrt{x^2 + 5}$.

Exercises and answers

Exercises 1 to 4

Solve the following equations by separating the variables:

(1) $\frac{dy}{dx} = \frac{x}{y}$ given that when $x = 2$, $y = -3$

(2) $\frac{dy}{dx} = \frac{3x}{2y - 1}$

(3) $xy\frac{dy}{dx} = \frac{x^3 - 1}{5 - 3y}$

(4) $x\frac{dy}{dx} = y - xy$ given that when $x = 1$, $y = 1$

Answer 1

If $\frac{dy}{dx} = \frac{x}{y}$ then $ydy = xdx$.

Integrating yields $\int ydy = \int xdx$. That is, $\frac{y^2}{2} = \frac{x^2}{2} + C$.

Now, when $x=2$, $y=-3$ so that $\frac{(-3)^2}{2}=\frac{(2)^2}{2}+C$ and so $C=\frac{5}{2}$, giving the final solution as $\frac{y^2}{2}=\frac{x^2}{2}+\frac{5}{2}$ or $y=\pm\sqrt{x^2+5}$.

Answer 2

If $\frac{dy}{dx}=\frac{3x}{2y-1}$ then $(2y-1)dy=3xdx$.

Integrating yields $\int(2y-1)dy=\int 3xdx$. That is, $y^2-y=\frac{3x^2}{2}+C$.

Answer 3

If $xy\frac{dy}{dx}=\frac{x^3-1}{5-3y}$ then $y(5y-3)dy=\left(\frac{x^3-1}{x}\right)dx$.

Integrating yields

$$\int y(5y-3)dy=\int\left(\frac{x^3-1}{x}\right)dx.$$

That is

$$\int(5y^2-3y)dy=\int\left(x^2-\frac{1}{x}\right)dx.$$

This integrates as

$$\frac{5y^3}{3}-\frac{3y^2}{2}=\frac{x^3}{3}-\ln x+C.$$

Answer 4

If $x\frac{dy}{dx}=y-xy$ then $\frac{dy}{y}=\left(\frac{1}{x}-1\right)dx$.

Integrating yields

$$\int\frac{dy}{y}=\int\left(\frac{1}{x}-1\right)dx.$$

That is, $\ln y=\ln x-x+C$. Now, when $x=1$, $y=1$ and so $\ln 1=\ln 1-1+C$ and since $\ln 1=0$ this means that $C=1$, giving the final solution as $\ln y=\ln x-x+1$ or $y=e^{\ln x-x+1}$.

Differential equations of the form $\frac{df(x)}{dx}+p(x)f(x)=0$ ***where*** $p(x)$ ***is known***

All equations of this type can be solved by using the method of separation of variables. $\frac{df(x)}{dx}+p(x)f(x)=0$ can be re-written as $\frac{df(x)}{f(x)}=-p(x)dx$. By integrating both sides we find that

$$\int\frac{df(x)}{f(x)}=-\int p(x)dx.$$

That is, $\ln f(x)=-\int p(x)dx$. In other words

$$f(x) = e^{-\int p(x)dx}.$$

For example, the solution of the differential equation $\frac{df(x)}{dx} + \sin x f(x) = 0$, where $f(0) = 5$, is given as

$$f(x) = e^{-\int \sin x dx} = e^{\cos x + C} = Ae^{\cos x}.$$

Since $f(0) = 5$ we see that $f(0) = Ae^{\cos 0} = 5$. Since $\cos 0 = 1$ and $e^1 = e$ we have that $Ae = 5$ and so $A = \frac{5}{e} = 5e^{-1}$ giving the final solution as $f(x) = 5e^{\cos x - 1}$.

Exercises and answers

Exercises 1 to 4

Solve each of the following differential equations:

(1) $\frac{df(x)}{dx} + xf(x) = 0$ where $f(0) = 2$

(2) $x\frac{df(x)}{dx} + f(x) = 0$ where $f(1) = 8$

(3) $\frac{df(x)}{dx} + \cos x f(x) = 0$ where $f(\pi/2) = 1$

(4) $\tan x\frac{df(x)}{dx} + f(x) = 0$ where $f(\pi/2) = -3$

Answer 1

If $\frac{df(x)}{dx} + xf(x) = 0$ then $f(x) = e^{-\int x dx} = e^{-\frac{x^2}{2} + C} = Ae^{-\frac{x^2}{2}}$.

Since $f(0) = 2$ we see that $f(0) = Ae^0 = 2$.

Since $e^0 = 1$ we have that $A = 2$, giving the final solution as

$$f(x) = 2e^{-\frac{x^2}{2}}$$

Answer 2

In order to put $x\frac{df(x)}{dx} + f(x) = 0$ into the usual form we must first divide throughout by x to give $\frac{df(x)}{dx} + \frac{1}{x}f(x) = 0$. Now we find that

$$f(x) = e^{-\int \frac{dx}{x}} = e^{-\ln x + C} = Ae^{-\ln x} = Ae^{\ln\left(\frac{1}{x}\right)} = \frac{A}{x}.$$

Since $f(1) = 8$ we see that $f(1) = \frac{A}{1} = 8$. This gives the final solution as $f(x) = \frac{8}{x}$.

Answer 3

If $\frac{df(x)}{dx} + \cos x f(x) = 0$ then $f(x) = e^{-\int \cos x dx} = e^{-\sin x + C} = Ae^{-\sin x}$. Since $f(\pi/2) = 1$ we see that $f(\pi/2) = Ae^{-\sin \pi/2} = Ae^{-1} = 1$. So we have that $A = e$, giving the final solution as $f(x) = e^{-\sin x + 1}$.

Answer 4

In order to put $\tan x \dfrac{df(x)}{dx} + f(x) = 0$ into the usual form we must first divide throughout by $\tan x$ to give $\dfrac{df(x)}{dx} + \dfrac{1}{\tan x} f(x) = 0$. Now we find that

$$f(x) = e^{-\int \frac{dx}{\tan x}} = e^{-\int \frac{\cos x dx}{\sin x}} = e^{-\ln \sin x + C} = Ae^{-\ln \sin x} = \frac{A}{\sin x}.$$

Since $f(\pi/2) = -3$ we see that $f(\pi/2) = \dfrac{A}{\sin \frac{\pi}{2}} = A = -3$. This gives the final solution as $f(x) = -\dfrac{3}{\sin x}$.

Equations of the form $\dfrac{df(x)}{dx} + p(x)f(x) = q(x)$ ***where*** $p(x)$ ***and*** $q(x)$ ***are known***

The solutions of equations of this type are given as

$$f(x) = \frac{1}{I(x)} \int I(x)q(x)dx,$$

where $I(x) = e^{\int p(x)dx}$.

For example, to solve the equation $\dfrac{df(x)}{dx} + \dfrac{f(x)}{x} = x^2$ given that $f(1) = 0$ we find that $f(x) = \dfrac{1}{I(x)} \int I(x)x^2 dx$ where $I(x) = e^{\int \frac{1}{x} dx} = e^{\ln x} = x$.

Notice that in evaluating the integral in the exponent we have ignored the integration constant. This is because it will appear in the subsequent integration that now follows. Using the form of the solution as above we see that

$$f(x) = \frac{1}{x} \int x.x^2 dx = \frac{1}{x} \int x^3 dx = \frac{1}{x}\left[\frac{x^4}{4} + C\right] = \frac{x^3}{4} + \frac{C}{x}.$$

Noting that $f(1) = 0$ we find that $f(1) = \dfrac{1}{4} + C = 0$ so $C = -\dfrac{1}{4}$, giving the final solution as $f(x) = \dfrac{x^3}{4} - \dfrac{1}{4x}$.

Exercises and answers

Exercises 1 to 4

Solve each of the following differential equations:

(1) $\dfrac{df(x)}{dx} + 3f(x) = e^x$ where $f(0) = 1$

(2) $x\dfrac{df(x)}{dx} + f(x) = x^2$ where $f(1) = -\dfrac{2}{3}$

(3) $x\dfrac{df(x)}{dx} - 4f(x) = x^5$ where $f(2) = 0$

(4) $\tan x \dfrac{df(x)}{dx} + f(x) = \dfrac{1}{\cos x}$ where $f(\pi/2) = 0$

Answer 1

If $\frac{df(x)}{dx} + 3f(x) = e^x$ then $f(x) = \frac{1}{I(x)}\int I(x)e^x dx$ where $I(x) = e^{\int 3dx} = e^{3x}$. Hence

$$f(x) = \frac{1}{e^{3x}}\int e^{3x}.e^x dx = \frac{1}{e^{3x}}\int e^{4x}dx = \frac{1}{e^{3x}}\left[\frac{e^{4x}}{4} + C\right] = \frac{e^x}{4} + \frac{C}{e^{3x}}$$

Noting that $f(0) = 1$ we find that $f(0) = \frac{1}{4} + C = 1$ so $C = \frac{3}{4}$, giving the final solution as

$$f(x) = \frac{e^x}{4} + \frac{3}{4e^{3x}} = \frac{1}{4}\left(e^x + 3e^{-3x}\right).$$

Answer 2

First rewrite $x\frac{df(x)}{dx} + f(x) = x^2$ in normal form by dividing throughout by x:

$$\frac{df(x)}{dx} + \frac{1}{x}f(x) = x.$$

Now $f(x) = \frac{1}{I(x)}\int I(x)xdx$, where $I(x) = e^{\int \frac{1}{x}dx} = e^{\ln x} = x$.

Hence $f(x) = \frac{1}{x}\int x.xdx = \frac{1}{x}\int x^2 dx = \frac{1}{x}\left(\frac{x^3}{3} + C\right)$.

Now $f(1) = -\frac{2}{3}$ and so $f(1) = -\frac{2}{3} = \frac{1}{3} + C$, giving $C = -1$ and the final solution as

$$f(x) = \frac{1}{x}\left(\frac{x^3}{3} - 1\right).$$

Answer 3

First rewrite $x\frac{df(x)}{dx} - 4f(x) = x^5$ in normal form by dividing throughout by x:

$$\frac{df(x)}{dx} - \frac{4}{x}f(x) = x^4.$$

Now $f(x) = \frac{1}{I(x)}\int I(x)x^4 dx$, where $I(x) = e^{-\int \frac{4}{x}dx} = e^{-4\ln x} = e^{\ln x^{-4}} = x^{-4}$.

Hence

$$f(x) = \frac{1}{x^{-4}}\int x^{-4}x^4 dx = \frac{1}{x^{-4}}\int dx = \frac{1}{x^{-4}}[x + C].$$

Now $f(2) = 0$ so that $C = -2$ and the final solution is $f(x) = x^4(x-2)$.

Answer 4

First we rewrite $\tan x\frac{df(x)}{dx} + f(x) = \frac{1}{\cos x}$ in normal form by dividing throughout

by $\tan x = \frac{\sin x}{\cos x}$:

$$\frac{df(x)}{dx} + \frac{\cos x}{\sin x}f(x) = \frac{1}{\sin x}.$$

Now, $f(x) = \frac{1}{I(x)} \int \frac{I(x)}{\sin x} dx$ where $I(x) = e^{\int \frac{\cos x}{\sin x} dx} = e^{\ln \sin x} = \sin x$.

Hence $f(x) = \frac{1}{\sin x} \int \frac{\sin x}{\sin x} dx = \frac{1}{\sin x} \int dx = \frac{x + C}{\sin x}$.

Now $f(\pi / 2) = 0$ so that $C = -\pi / 2$ and the final solution is

$$f(x) = \frac{x - \pi / 2}{\sin x}.$$

H2 DIFFERENCE EQUATIONS

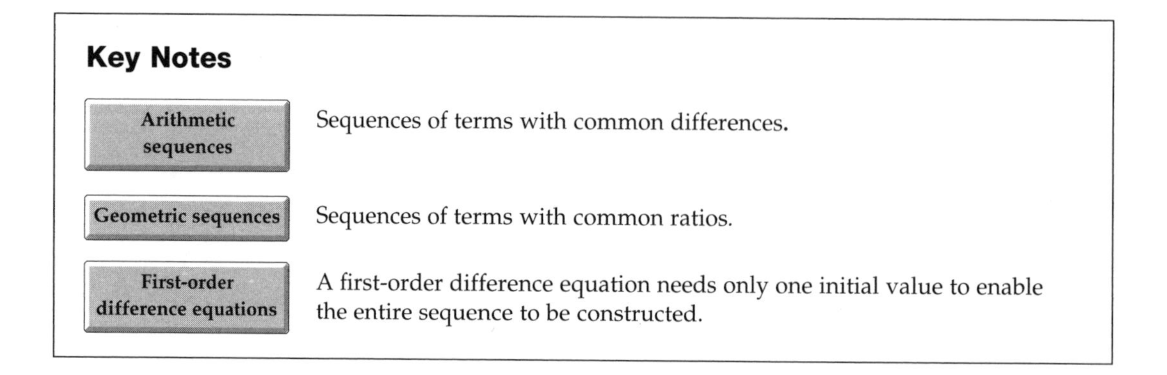

Key Notes

Arithmetic sequences	Sequences of terms with common differences.
Geometric sequences	Sequences of terms with common ratios.
First-order difference equations	A first-order difference equation needs only one initial value to enable the entire sequence to be constructed.

Arithmetic sequences

The sequence of numbers

$$2, 5, 8, 11, 14, \ldots$$

form what is called an **arithmetic sequence**. The sequence begins with the number 2 – we call this the **zeroth term** – and then each term differs from its predecessor by the same amount, namely 3 – we call this the **common difference**. The ellipsis … indicates that the sequence continues without end. We have a general form for all arithmetic sequences, namely:

$$a_n = a_0 + nd.$$

Here a_n stands for the nth term of the sequence, a_0 stands for the zeroth term and d stands for the common difference. For the specific sequence above, $a_0 = 2$ and $d = 3$ so that

$$a_n = 2 + 3n.$$

This formula is sufficient to generate any term in the arithmetic sequence. An alternative way of describing this sequence is to say that the $(n+1)^{\text{th}}$ term of the sequence is equal to the nth term of the sequence plus 3. That is

$$a_{n+1} = a_n + 3.$$

An equation of this type is called a **difference equation** where the general term of the sequence is given in terms of a predecessor. This, however, is not sufficient alone to generate the sequence. We need to know the value of a_0 to start it off. In this case $a_0 = 2$.

Geometric sequences

The sequence of numbers

$$1, 4, 16, 64, 256, \ldots$$

form what is called a **geometric sequence**. The sequence begins with the number 1 – we call this the **zeroth term** and then the ratio of each term to its predecessor is the same, namely 4 – we call this the **common ratio**. We have a general form for all geometric sequences, namely:

$$a_n = a_0 r^n.$$

Here a_n stands for the nth term of the sequence, a_0 stands for the zeroth term and r stands for the common ratio. For the specific sequence above, $a_0 = 1$ and $r = 4$ so that

$$a_n = 1 \times 4^n = 4^n.$$

This formula is sufficient to generate any term in the geometric sequence. As with the arithmetic sequence we can describe the geometric sequence as a difference equation the $(n+1)^{th}$ term of the sequence is equal to the nth term of the sequence multiplied by 4. That is:

$$a_{n+1} = 4a_n.$$

Again, however, this is not sufficient alone to generate the sequence. We need to know the value of a_0 to start it off. In this case $a_0 = 1$.

Exercises and answers

Exercises 1 to 5
Write each of the following series as a difference equation:

(1) 1, 6, 11, 16, 21, 26, …
(2) 9, 6, 3, 0, –3, –6, …
(3) 1, 3, 9, 27, 81, …
(4) 100, 10, 1, 0.1, 0.01, 0.001, …
(5) 4, –8, 16, –32, 64, …

Answer 1
This is an arithmetic series because the common difference (d) between any two successive terms is the same, namely 5. The zeroth term (a_0) is 1 so the nth term is given as

$$a_n = 1 + 5n,$$

which is written as a difference equation in the form

$$a_{n+1} = a_n + 5, \text{ where } a_0 = 1.$$

Answer 2
This is an arithmetic series because the common difference (d) between any two successive terms is the same, namely –3. The zeroth term (a_0) is 9 so the nth term is given as

$$a_n = 9 - 3n,$$

which is written as a difference equation in the form

$$a_{n+1} = a_n - 3, \text{ where } a_0 = 9.$$

Answer 3
This is a geometric series because the common ratio (r) between any two successive terms is the same, namely 3. The zeroth term (a_0) is 1 so the nth term is given as

$$a_n = 3^n,$$

which is written as a difference equation in the form

$$a_{n+1} = 3a_n, \text{ where } a_0 = 1.$$

Answer 4

This is a geometric series because the common ratio (r) between any two successive terms is the same, namely 0.1. The zeroth term (a_0) is 100 so the nth term is given as

$$a_n = 100 \times 0.1^n,$$

which is written as a difference equation in the form

$$a_{n+1} = 0.1a_n, \text{ where } a_0 = 100.$$

Answer 5

This is a geometric series because the common ratio (r) between any two successive terms is the same, namely –2. The zeroth term (a_0) is 4 so the nth term is given as

$$a_n = 4 \times (-2)^n,$$

which is written as a difference equation in the form

$$a_{n+1} = -2a_n, \text{ where } a_0 = 4.$$

First-order difference equations

A **first-order difference equation** needs just one initial value to enable the entire sequence to be constructed. For example, if we start with the equation

$$a_{n+1} = \frac{n+1}{4}a_n$$

we can substitute $n - 1$ for n to give another equation:

$$a_n = \frac{n}{4}a_{n-1}.$$

Substituting this result into the first equation gives

$$a_{n+1} = \frac{n+1}{4} \times \frac{n}{4}a_{n-1}.$$

Indeed, by substituting $n - 2$ for n in the first equation we find that

$$a_{n-1} = \frac{n-1}{4}a_{n-2} \text{ and so } a_{n+1} = \frac{n+1}{4} \times \frac{n}{4} \times \frac{n-1}{4}a_{n-2}.$$

This pattern is repeated to give

$$a_{n+1} = \frac{n+1}{4} \times \frac{n}{4} \times \frac{n-1}{4} \times \frac{n-2}{4}a_{n-3} = \frac{n+1}{4} \times \frac{n}{4} \times \frac{n-1}{4} \times \frac{n-2}{4} \times \frac{n-3}{4}a_{n-4} = \cdots.$$

Eventually, we find that

$$a_{n+1} = \frac{n+1}{4} \times \frac{n}{4} \times \frac{n-1}{4} \times \frac{n-2}{4} \times \frac{n-3}{4} \times \cdots \times \frac{1}{4}a_{n-n}.$$

That is:

$$a_{n+1} = \frac{(n+1)!}{4^{n+1}}a_0$$

so that any term of the sequence can be found once we know the value of a_0. If, for example $a_0 = 3$ then $a_{n+1} = \frac{3(n+1)!}{4^{n+1}}$.

Exercises and answers

Exercises 1 to 5

Find the next four terms of each of the following sequences:

(1) $a_{n+1} = \frac{(n+1)!}{2^n} a_0$ where $a_0 = 5$

(2) $a_{n+1} = (-1)^n a_0$ where $a_0 = 1$

(3) $a_{n+1} = (2n-7) a_0$ where $a_0 = 1$

(4) $a_{n+1} = \frac{n}{2}(n+1)a_0$ where $a_0 = -4$

(5) $a_{n+1} = 7^{n+2} a_n$ where $a_0 = 1$

Answer 1

Here

$$\begin{aligned} a_1 &= \frac{(1)!}{2^0} a_0 = 5 \\ a_2 &= \frac{(2)!}{2^1} a_0 = \frac{2}{2} \times 5 = 5 \\ a_3 &= \frac{(3)!}{2^2} a_0 = \frac{6}{4} \times 5 = \frac{15}{2} \\ a_4 &= \frac{(4)!}{2^3} a_0 = \frac{24}{8} \times 5 = 15 \end{aligned}$$

Answer 2

Here

$$\begin{aligned} a_1 &= (-1)^0 a_0 = 1 \times 1 = 1 \\ a_2 &= (-1)^1 a_0 = (-1) \times 1 = -1 \\ a_3 &= (-1)^2 a_0 = 1 \times 1 = 1 \\ a_4 &= (-1)^3 a_0 = (-1) \times 1 = -1 \end{aligned}$$

Answer 3

Here

$$\begin{aligned} a_1 &= (2 \times 0 - 7) a_0 = -7 \times 1 = -7 \\ a_2 &= (2 \times 1 - 7) a_0 = -5 \times 1 = -5 \\ a_3 &= (2 \times 2 - 7) a_0 = -3 \times 1 = -3 \\ a_4 &= (2 \times 3 - 7) a_0 = -1 \times 1 = -1 \end{aligned}$$

Answer 4
Here

$$\begin{aligned} a_1 &= \frac{0}{2}(0+1)a_0 = 0 \\ a_2 &= \frac{1}{2}(1+1)a_0 = 1\times(-4) = -4 \\ a_3 &= \frac{2}{2}(2+1)a_0 = 3\times(-4) = -12 \\ a_4 &= \frac{3}{2}(3+1)a_0 = 6\times(-4) = -24 \end{aligned}$$

Answer 5
Here

$$\begin{aligned} a_1 &= 7^2 a_0 = 49\times 1 = 49 \\ a_2 &= 7^3 a_1 = 343\times 49 = 16,807 \\ a_3 &= 7^4 a_2 = 2401\times 16,807 = 40,353,607 \\ a_4 &= 7^5 a_3 = 16,807\times 40,353,607 = 678,223,072,849 \end{aligned}$$

Solutions of difference equations
The equation

$$a_n = 3n(n-1)$$

satisfies the difference equation

$$a_n - a_{n-1} = 6(n-1)$$

because

$$\begin{aligned} a_n - a_{n-1} &= 3n(n-1) - 3(n-1)(n-2) \\ &= 3n^2 - 3n - 3(n^2 - 3n + 2) \\ &= 3n^2 - 3n - 3n^2 + 9n - 6 \\ &= 6n - 6 \\ &= 6(n-1). \end{aligned}$$

We say that $a_n = 3n(n-1)$ is a solution of the difference equation $a_n - a_{n-1} = 6(n-1)$.

Exercises and answers

Exercises 1 to 3
Show that

(1) $a_n = 4n - 5$ is a solution of $a_n - a_{n-1} = 4$
(2) $x_n = n$ is a solution of $3x_n + 4x_{n-1} = 7n - 4$
(3) $p_n = n^2$ is a solution of $p_{n+1} - p_{n-1} = 4n$

Answer 1
If $a_n = 4n - 5$ then

$$\begin{aligned} a_n - a_{n-1} &= (4n-5) - (4[n-1]-5) \\ &= 4n - 5 - (4n - 9) \\ &= 4n - 5 - 4n + 9 \\ &= 4 \text{ as required.} \end{aligned}$$

Answer 2
If $x_n = n$ then

$$\begin{aligned} 3x_n + 4x_{n-1} &= 3n + 4(n-1) \\ &= 7n - 4 \text{ as required.} \end{aligned}$$

Answer 3
If $p_n = n^2$ then

$$\begin{aligned} p_{n+1} - p_{n-1} &= (n+1)^2 - (n-1)^2 \\ &= (n^2 + 2n + 1) - (n^2 - 2n + 1) \\ &= n^2 + 2n + 1 - n^2 + 2n - 1 \\ &= 4n \text{ as required.} \end{aligned}$$

I1 POPULATION GROWTH

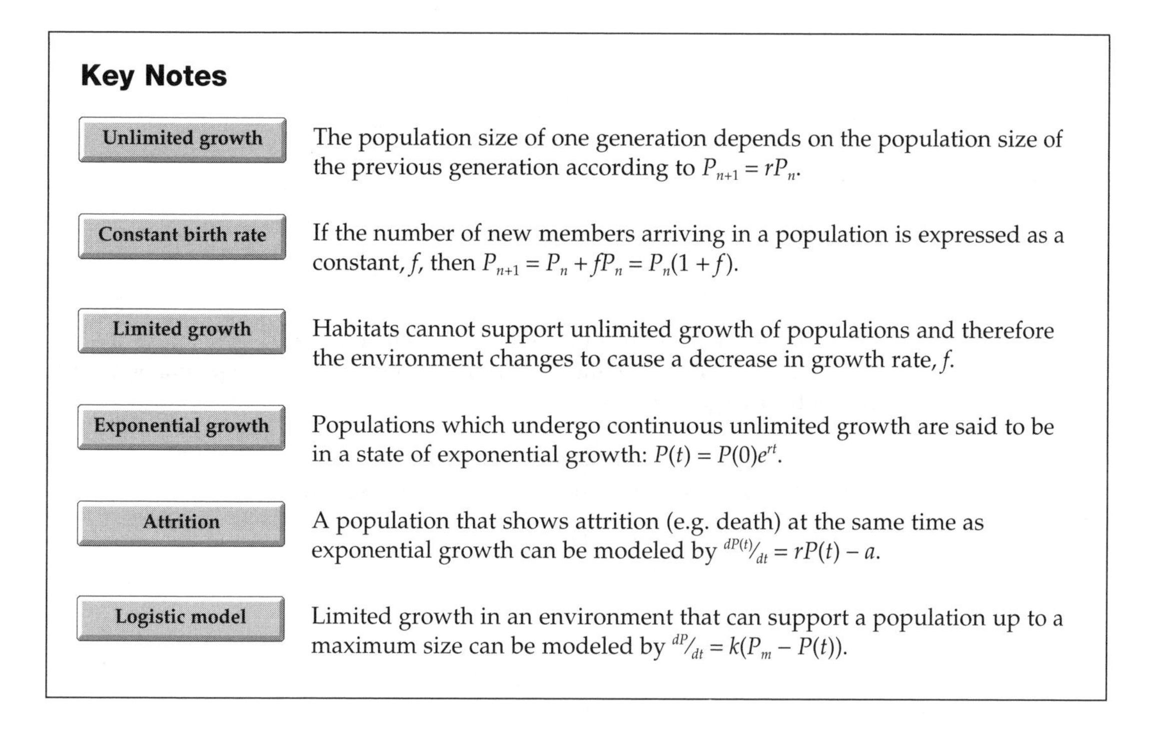

Key Notes

Unlimited growth	The population size of one generation depends on the population size of the previous generation according to $P_{n+1} = rP_n$.
Constant birth rate	If the number of new members arriving in a population is expressed as a constant, f, then $P_{n+1} = P_n + fP_n = P_n(1 + f)$.
Limited growth	Habitats cannot support unlimited growth of populations and therefore the environment changes to cause a decrease in growth rate, f.
Exponential growth	Populations which undergo continuous unlimited growth are said to be in a state of exponential growth: $P(t) = P(0)e^{rt}$.
Attrition	A population that shows attrition (e.g. death) at the same time as exponential growth can be modeled by ${}^{dP(t)}/_{dt} = rP(t) - a$.
Logistic model	Limited growth in an environment that can support a population up to a maximum size can be modeled by ${}^{dP}/_{dt} = k(P_m - P(t))$.

Unlimited growth

The simplest reason for population growth is where the population size, P, for one generation depends upon the population size of the previous generation.

$$P_{n+1} = rP_n.$$

where r is called the Malthusian factor that determines the rate of growth and n is the generation number.

Exercises and answers

Exercise 1

Show that for unlimited growth the population in generation n is r^n multiplied by the initial population.

Exercise 2

Find the difference equation for the population growth if the difference in growth of a population in two successive generations is always in the same constant proportion to the population in the first of the two generations.

Exercise 3

How does the population in generation n relate to the initial population in the population model of question 2.

Exercise 4
An animal population is increasing by 5% each year. If the initial population is 500 after how many years will it exceed 750?

Answer 1
As $P_{n+1} = rP_n$ then

$$P_1 = rP_0$$
$$P_2 = rP_1 = rrP_0 = r^2P_0$$
$$P_3 = rP_2 = rr^2P_0 = r^3P_0$$
$$\cdots\cdots\cdots$$
$$P_n = r.r^{n-1}P_0 = r^nP_0$$

Answer 2
The growth of a population between generation n and generation $n+1$ is $P_{n+1} - P_n$. This is always in the same constant proportion to the population in the first of the two generations. So that

$$P_{n+1} - P_n = kP_n$$

where k is the constant of proportionality. This means that

$$P_{n+1} = P_n + kP_n = (1+k)P_n.$$

Answer 3
As $P_{n+1} = (1+k)P_n$ then

$$P_1 = (1+k)P_0$$
$$P_2 = (1+k)P_1 = (1+k)^2P_0$$
$$P_3 = (1+k)P_2 = (1+k)^3P_0$$
$$\cdots\cdots\cdots$$
$$P_n = (1+k)P_{n-1} = (1+k)^nP_0.$$

Answer 4
An animal population is increasing by 5% each year so that $P_{n+1} = rP_n$ where $r = 1.05$ (100% + 5%). If $P_0 = 500$ this means that

$$P_n = r^nP_0 = (1.05)^n 500.$$

If $P_n = 750$ then $750 = (1.05)^n 500$. This means that

$$\frac{750}{500} = (1.05)^n \text{ so that } (1.05)^n = 1.5.$$

Taking the logarithm of each side yields

$$n\ln(1.05) = \ln(1.5)$$

and so

$$n = \frac{\ln(1.5)}{\ln(1.05)} = 8.3$$

to one decimal place. This means that the population will exceed 750 during the 8th year.

Constant birth rate

If the number of new members arriving in a population can be expressed as a constant f per member of the population then

$$P_{n+1} = P_n + fP_n = P_n(1+f).$$

Since this formula can be applied year on year we can see that

$$P_{n+1} = P_n(1+f) = P_{n-1}(1+f)^2 = P_{n-2}(1+f)^3 = \ldots = P_0(1+f)^n$$

where P_0 is the initial population.

Exercises and answers

Exercise 1

A population of 1 000 grows by 15% each year. Write down a table of values for the population size for the first 5 years.

Exercise 2

The natural growth rate of a stock of fish in the North Sea is 25%. Fishermen are permitted to harvest 500 000 fish per year. If P_n denotes the number of fish, in millions, find a difference equation that describes the change in population year on year.

Exercise 3

How long will it take to reduce the fish stock size in Exercise 2 to zero if the initial stock size is

(1) 1 million
(2) 10 million?

Exercise 4

What is the initial stock size in Exercise 2 that maintains equilibrium in the population?

Answer 1

A population of 1000 grows by 15% each year and so

$$P_{n+1} = P_0(1+f)^n$$

where $P_0 = 1\,000$ (the initial population) and $f = 0.15$ (the 15% increase). Consequently,

$$P_{n+1} = 1\,000(1.15)^n$$

with the values given in *Table 1*.

Table 1. The 5 year increase in an initial population of 1 000, growing at 15% per year

Year n	Population P_n
1	$P_1 = 1000(1.15) = 1150$
2	$P_2 = 1000(1.15)^2 = 1323$
3	$P_3 = 1000(1.15)^3 = 1521$
4	$P_4 = 1000(1.15)^4 = 1749$
5	$P_5 = 1000(1.15)^5 = 2011$

Answer 2

The natural growth rate of a stock of fish in the North Sea is 25% so if P_n represents the stock size in year n (measured in millions) then

increase in population = natural growth – harvested quantity:

$$P_n = (1.25)P_{n-1} - 0.5$$
$$P_n = (1.25)P_{n-1} - 0.5 \text{ (millions).}$$

Answer 3

Reducing the fish stock will take the following times.

(1) $P_0 = 1$ (measured in millions) and so:

$$P_1 = (1.25)P_0 - 0.5 = 1.25 - 0.5 = 0.75$$
$$P_2 = (1.25)P_1 - 0.5 = 1.25 \times 0.75 - 0.5 = 0.437\ 5$$
$$P_3 = (1.25)P_2 - 0.5 = 1.25 \times 0.437\ 5 - 0.5 = 0.046\ 9$$
$$P_4 = (1.25)P_3 - 0.5 = 1.25 \times 0.046\ 9 - 0.5 = -0.441\ 4.$$

Naturally, the population can never be negative so what this is saying is that the population will be wiped out during the 4th year of fishing.

(2) $P_0 = 10$ (measured in millions) and so

$$P_1 = (1.25)P_0 - 0.5 = 12.5 - 0.5 = 12.0$$
$$P_2 = (1.25)P_1 - 0.5 = 1.25 \times 12.0 - 0.5 = 14.5.$$

It is evident that on this rate of harvesting the fish stock size will increase and never become zero on this model.

Answer 4

The initial stock size that maintains equilibrium in the population can be found from

$$P_n - P_{n-1} = (0.25)P_{n-1} - 0.5.$$

If equilibrium is maintained then $P_n - P_{n-1} = 0$ so that $(0.25)P_{n-1} - 0.5 = 0$ where P_{n-1} is a constant value and equal to P_0. This means that

$$(0.25)P_0 - 0.5 = 0,$$

so that

$$P_0 = \frac{0.5}{0.25} = 2.$$

So an initial stock size of 2 million fish could sustain this harvest and maintain population equilibrium.

Limited growth

The population described by the difference equation $P_{n+1} = rP_n$ or $P_{n+1} = (1+f)P_n$ will continue to grow indefinitely. Clearly, no habitat can support such growth as the capacity to support the population will be insufficient to maintain the growth. What happens is that as the population increases, the growth rate, f, decreases: the growth rate decreases as a consequence of the change in the environment. To take account of this in our model we take the simplest growth rate that is itself linearly dependent upon the current population. This is a linear form:

$$f = a - bP_n.$$

Substituting this into the original difference equation gives

$$P_{n+1} = (1 + a - bP_n)P_n = (r - bP_n)P_n = r\left(1 - \frac{b}{r}P_n\right)P_n$$

The value of $\frac{a}{b}$ is the **equilibrium population** – the stable population level that can be supported by the environment.

Letting $C = \frac{r}{b}$ this equation becomes

$$P_{n+1} = r\left(1 - \frac{1}{C}P_n\right)P_n.$$

The quantity C is called the **carrying capacity**. In such a situation the growth of the population is moderated by the habitat so that when the population size is close to C the rate of growth is decreased and when it is much less than C the growth is increased. This equation is called the **logistic difference** equation.

Exercises and answers

Exercise 1

A population has a birth rate of $0.8 - 0.00004P_n$ and a death rate of $0.5 + 0.0002P_n$. Construct a table of values for the population for successive years with an initial population of 1 000.

Exercise 2

Repeat Exercise 1 with an initial population of 2 000.

Exercise 3

Investigate the behavior of the population described by $P_{n+1} = (2 - 0.002P_n)P_n$, for the initial populations:

(1) $P_0 = 800$
(2) $P_0 = 1000$.

Exercise 4

Investigate the behavior of the population described by $P_{n+1} = (3.1 - 0.001P_n)P_n$, for the initial population of $P_0 = 3095$.

Answer 1

The number of births between year n and year $n + 1$ will be

$$(0.8 - 0.000\,04P_n)P_n.$$

The number of deaths between year n and year $n + 1$ will be

$$(0.5 + 0.000\,2P_n)P_n.$$

This means that the population change $P_{n+1} - P_n$ will be given as

$$\begin{aligned} P_{n+1} - P_n &= (0.8 - 0.000\,04P_n)P_n - (0.5 + 0.000\,2P_n)P_n \\ &= (0.3 - 0.000\,24P_n)P_n. \end{aligned}$$

Notice that

$$\frac{0.3}{0.000\,24} = 1\,250$$

is the **equilibrium population**. Clearly,

$$P_{n+1} = (1.3 - 0.000\,24P_n)P_n$$

Starting from an initial population of $P_0 = 1\,000$ we find that

$$\begin{aligned} P_1 &= (1.3 - 0.000\,24P_0)P_0 \\ &= (1.3 - 0.24)1000 = 1060. \\ P_2 &= (1.3 - 0.000\,24P_1)P_1 \\ &= 1108. \\ P_3 &= (1.3 - 0.000\,24P_2)P_2 \\ &= 1146. \end{aligned}$$

Eventually, we obtain the values given in Table 2. As you can see from the table the population is approaching the equilibrium level of 1 250.

Answer 2

Starting from an initial population of $P_0 = 2000$ we find that

$$\begin{aligned} P_1 &= (1.3 - 0.000\,24P_0)P_0 \\ &= (1.3 - 0.48)2\,000 = 1\,640. \\ P_2 &= (1.3 - 0.000\,24P_1)P_1 \\ &= 1\,486. \end{aligned}$$

Now we see that the population is decreasing. Eventually, we obtain the values given in *Table 3*. Again, as you can see from the table the population is approaching the equilibrium level of 1 250.

Table 2. The population for successive years with an initial population of 1 000

Year *n*	P_n	Year *n*	P_n
0	1000	9	1236
1	1060	10	1240
2	1108	11	1243
3	1146	12	1245
4	1175	13	1247
5	1196	14	1248
6	1211	15	1248
7	1223	16	1249
8	1231		

The birth rate is given by $0.8 - 0.000\,04P_n$ and the death rate is given by $0.5 + 0.000\,2P_n$.

Table 3. The population for successive years with an initial population of 2 000

Year *n*	P_n	Year *n*	P_n
0	2000	9	1265
1	1640	10	1261
2	1486	11	1258
3	1402	12	1255
4	1351	13	1254
5	1318	14	1253
6	1297	15	1252
7	1282	16	1251
8	1272		

The birth rate is given by $0.8 - 0.000\,04P_n$ and the death rate is given by $0.5 + 0.000\,2P_n$.

Answer 3
As the population is described by $P_{n+1} = (2 - 0.002P_n)P_n$, then

(1) For the initial population level of $P_0 = 800$ produces the results given in *Table 4*. Here the population levels drop and then rise to the equilibrium value of 500.
(2) For the initial population level of $P_0 = 1000$ the results become those given in *Table 5*. The population is immediately extinguished.

Answer 4
The population described by $P_{n+1} = (3.1 - 0.001P_n)P_n$ with the initial population of $P_0 = 3095$ produces the values given in *Table 6*. The population becomes unstable. Indeed, if any population for which $a > 2$ the population becomes unstable whenever it approaches the equilibrium level. In this problem the equilibrium population is 3100.

Exponential growth

If, instead of thinking in terms of population growths every generation as though the population suddenly changes at the end of each generation we think of a continuous growth in time then for large populations we shall be nearer the actual state of affairs. Again, unlimited growth is our starting point which for the continuous growth case is referred to as **exponential growth**.

If we assume that the rate at which the population changes is proportional to the existing population, that is,

Table 4. Change of an initial population of 800 over 6 years

Year *n*	P_n	Year *n*	P_n
0	800	4	500
1	320	5	500
2	435	6	500
3	492		

The change is described by $P_{n+1} = (2 - 0.002P_n)P_n$.

Table 5. Change of an initial population of 1000 over 6 years

Year *n*	P_n	Year *n*	P_n
0	1000	4	0
1	0	5	0
2	0	6	0
3	0		

The change is described by $P_{n+1} = (2 - 0.002P_n)P_n$.

Table 6. Change of an initial population of 800 for a population described by $P_n + 1 = (3.1 - 0.001P_n)P_n$

Year *n*	P_n	Year *n*	P_n
0	3095	7	1924
1	15	8	2263
2	48	9	1895
3	146	10	2284
4	430	11	1864
5	1149	12	2304
6	2242	13	1834

$$\frac{dP(t)}{dt} \propto P(t)$$

because $\frac{dP(t)}{dt}$ represents the rate of change of $P(t)$ with respect to time t. This gives us that

$$\frac{dP(t)}{dt} = rP(t)$$

where r is the proportionality constant – the **growth factor**.

To solve this equation we employ the method of separation of variables:

$$\frac{dP(t)}{P(t)} = rdt$$

which we now integrate

$$\int \frac{dP(t)}{P(t)} = \int rdt$$

with the result

$$\ln P(t) = rt + C,$$

that is, $P(t) = e^{rt+C} = Ae^{rt}$ where $A = e^C$. Now, when $t = 0$ this becomes $P(0) = Ae^0 = A$ so the final solution is

$$P(t) = P(0)e^{rt}.$$

Exercise

A population has a growth factor of 20%. How long will it take for the population to double?

Answer

Let $P(0)$ be the initial population so that when $P(t) = P(0)e^{rt} = 2P(0)$ we see that $e^{rt} = 2$. Now, $r = \frac{20}{100} = 0.2$ so that $e^{0.2t} = 2$ and by taking logarithmns of both sides we see that $0.2t = \ln 2$ so that $t = 5\ln 2 = 3.47$. Consequently, if the growth factor is measured as 20% of the population per year then it will take 3.47 years to double. If the growth factor is measured as 20% of the population per day then it will take 3.47 days to double.

Attrition

If a population has an exponential growth but at the same time an attrition, for example by death or by emigration, then this can be modeled by

$$\frac{dP(t)}{dt} = rP(t) - a.$$

If the population is changing as a consequence of immigration then the value of will be negative. Manipulating this equation we find that

$$\frac{dP(t)}{dt} = r\left(P(t) - \frac{a}{r}\right)$$

and so, separating the variables,

$$\frac{dP(t)}{\left(P(t) - \frac{a}{r}\right)} = rdt.$$

Integration yields

$$\int \frac{dP(t)}{\left(P(t) - \frac{a}{r}\right)} = \int r dt$$

that is,

$$\ln\left(P(t) - \frac{a}{r}\right) = rt + C.$$

That is,

$$P(t) - \frac{a}{r} = e^{rt+C} = Ae^{rt}$$

Where

$$P(0) - \frac{a}{r} = A.$$

This gives the final solution as

$$P(t) - \frac{a}{r} = \left(P(0) - \frac{a}{r}\right)e^{rt} \text{ or } P(t) = \left(P(0) - \frac{a}{r}\right)e^{rt} + \frac{a}{r}.$$

Exercises and answers

Exercise 1
A population has an initial size of 1 000 and a growth factor of 4%. If the population is decreasing at a rate of 100 per year what will be the size of the population after:

(1) 5 years
(2) 10 years?

Exercise 2
A population has an initial size of 1 000 and a depletion factor of 4%. If the population is increasing at a rate of 100 per year what will be the size of the population after:

(1) 5 years
(2) 10 years?

Answer 1
Use

$$P(t) = \left(P(0) - \frac{a}{r}\right)e^{rt} + \frac{a}{r}$$

where $P(0) = 1000,\ r = 0.04$ and $a = 100$ so that

$$P(t) = \left(1000 - \frac{100}{0.04}\right)e^{0.04t} + \frac{100}{0.04}$$

and so after 5 years

$$P(5) = \left(1000 - \frac{100}{0.04}\right)e^{0.2} + \frac{100}{0.04} = 668,$$

and after 10 years

$$P(10) = \left(1000 - \frac{100}{0.04}\right)e^{0.4} + \frac{100}{0.04} = 262.$$

Thus, the population is decreasing year on year.

Answer 2
Use

$$P(t) = \left(P(0) - \frac{a}{r}\right)e^{rt} + \frac{a}{r}$$

where $P(0) = 1000$, $r = -0.04$ and $a = -100$ so that,

$$P(t) = \left(1000 - \frac{100}{0.04}\right)e^{-0.04t} + \frac{100}{0.04}.$$

So after 5 years

$$P(5) = \left(1000 - \frac{100}{0.04}\right)e^{-0.2} + \frac{100}{0.04} = 1272,$$

and after 10 years

$$P(10) = \left(1000 - \frac{100}{0.04}\right)e^{-0.4} + \frac{100}{0.04} = 1495.$$

Thus, the population is increasing year on year.

Logistic model

Limited growth, as exemplified by an environment that can support a population but only up to a maximum size, can be simply modeled by

$$\frac{dP(t)}{dt} = k(P_m - P(t))$$

where P_m is the maximum supportable population and k is a constant. Clearly, as the population approaches the maximum then the rate of change slows down. This differential equation becomes identical to the previous one studied when $kP_m = -a$ and $k = -r$. This gives the solution to this equation as

$$P(t) = P(0)e^{-kt} + P_m(1 - e^{-kt}).$$

Again, as in the last model, as time increases the population increases or decreases to the equilibrium level of P_m. The limitations of this model are evident in that when the population tends to zero the model approximates to

$$\frac{dP(t)}{dt} = kP_m$$

whereas, in fact, the rate of change should tend to zero as well.

An improved model that takes account of this point is the **logistic model**. Again, this is the limited growth expressed as a differential equation:

$$\frac{dP(t)}{dt} = kP(t)(P_m - P(t)).$$

The solution to this equation can be obtained by separating the variables. The expression

$$\frac{dP(t)}{dt} = kP(t)\left(P_m - P(t)\right)$$

becomes

$$\frac{dP(t)}{P(t)\left(P_m - P(t)\right)} = kdt.$$

For convenience we write $x = P(t)$ and $x_m = P_m$ so that the equation can be written as

$$\frac{dx}{x\left(x_m - x\right)} = kdt.$$

Now, by partial fractions

$$\frac{1}{x\left(x_m - x\right)} = \frac{1}{x_m}\left\{\frac{1}{x} + \frac{1}{x_m - x}\right\}$$

and so

$$\int \frac{dx}{x\left(x_m - x\right)} = \frac{1}{x_m}\int\left\{\frac{1}{x} + \frac{1}{x_m - x}\right\}dx = \int kdt.$$

That is,

$$\int\left\{\frac{1}{x} + \frac{1}{x_m - x}\right\}dx = -\int adt,$$

by multiplying by $x_m = P_m$ and noting that $kP_m = -a$. The solution is then

$$\ln x - \ln(x_m - x) = \ln\frac{x}{x_m - x} = -at + C,$$

that is,

$$\frac{x}{x_m - x} = e^{-at+C}.$$

That is,

$$\frac{P(t)}{P_m - P(t)} = e^{-at+C}.$$

Clearly,

$$\frac{P(0)}{P_m - P(0)} = e^{C}.$$

Now, this means that

$$P(t) = \left(P_m - P(t)\right)e^{-at+C} = \left(P_m - P(t)\right)\frac{P(0)}{P_m - P(0)}e^{-at}$$

and so

$$P(t) = \frac{P_m P(0)}{P(0) + (P_m - P(0))e^{at}}$$

where $a = -kP_m$. This is the solution to the logistic equation.

Exercise
An environment that can support a population of 10 000 suddenly changes to one that can only support 5 000. How long will it take the population to fall to 6 000 if $a = -0.02$ per day?

Answer
Use

$$P(t) = \frac{P_m P(0)}{P(0) + (P_m - P(0))e^{at}}$$

where $P_m = 5\,000$, $P(0) = 10\,000$ and $P(t) = 6\,000$ so that

$$6\,000 = \frac{5\,000 \times 10\,000}{10\,000 + (5\,000 - 10\,000)e^{-0.02t}};$$

that is,

$$6\,000 \times 10\,000 - 6\,000 \times 5\,000e^{-0.02t} = 5\,000 \times 10\,000$$

so that $1\,000 \times 10\,000 = 6\,000 \times 5\,000e^{-0.02t}$, giving

$$e^{-0.02t} = \frac{1}{3}$$

and so

$$-0.02t = \ln\left(\frac{1}{3}\right).$$

Consequently,

$$t = -\frac{1}{0.02}\ln\left(\frac{1}{3}\right) \cong 55 \text{ days.}$$

12 HEAT LOSS FROM A BODY

Key Note

Newtons' law of cooling

The rate at which a body loses heat is proportional to the difference in temperature between the body and its surroundings. It is an empirical law that is only true for substantial temperature differences if the heat loss is by convection or conduction.

Newton's law of cooling

If you blow on a hot cup of coffee the air moving over the surface of the coffee, being at a lower temperature, will **convect** energy away from the coffee and cause it to cool down. The rate at which cooling occurs is directly proportional to the temperature difference between the coffee and the forced, moving draught. This is the essence of Newton's law of cooling and it can be expressed in the form of a differential equation.

If $T(t)$ represents the temperature of a body at time t and T_e is the ambient temperature of the environment (maintained by a forced draught) then

$$\frac{dT(t)}{dt} \propto (T(t) - T_e).$$

Consequently,

$$\frac{dT(t)}{dt} = -k(T(t) - T_e)$$

where k is a positive proportionality constant. Since the temperature of the body is higher than the temperature of the environment we see that $T(t) - T_e$ is a positive number. Also, since the temperature of the body is decreasing – it is cooling down – the rate of change of temperature is negative. That is

$$\frac{dT(t)}{dt} < 0.$$

Consequently, the proportionality constant k is a positive number that depends upon the surface properties of the material being cooled.

To find the temperature $T(t)$ of the body at some time t requires us to solve this differential equation. This can be done quite straightforwardly by making the substitution

$$y(t) = T(t) - T_e \text{ so that } \frac{dy(t)}{dt} = \frac{dT(t)}{dt}.$$

The differential equation then becomes

$$\frac{dy(t)}{dt} = -ky(t)$$

with the familiar solution $y(t) = y(0)e^{-kt}$. This means that the solution to Newton's equation is

$$T(t) - T_e = (T(0) - T_e)e^{-kt},$$

that is

$$T(t) = T_e + (T(0) - T_e)e^{-kt}.$$

Exercise

A hot body has a temperature difference of 60°C above its surroundings. After 10 min this difference is 30°C. What will be the temperature difference after 20 min and how long will it take until the temperature difference is only 10°C?

Answer

Given that

$$T(t) = T_e + (T(0) - T_e)e^{-kt}$$

where $(T(0) - T_e) = 60$, then when $t = 10$ we have that $T(10) - T_e = 30$. That is,

$$T(10) - T_e = (T(0) - T_e)e^{-kt}$$

so that $30 = 60e^{-10k}$, giving $e^{-10k} = 0.5$. Taking logs of both sides yields

$$-10k = \ln 0.5 = -0.6931,$$

so that

$$k = 0.06931.$$

Consequently,

$$T(t) - T_e = 60e^{-0.06931t},$$

so that when $t = 20$, the temperature difference is

$$(T(20) - T_e = 60e^{-0.06931 \times 20})°C.$$
$$= 15°C$$

Furthermore, when $T(t) - T_e = 10$ then $10 = 60e^{-0.06931t}$, giving

$$e^{-0.06931t} = 0.1667.$$

Again, taking logs gives $-0.06931t = -1.7918$, and so $t = 25.9$ min.

13 CHEMICAL KINETICS

Key Notes

Radioactivity

The spontaneous disintegration of certain atomic nuclei to alpha and beta particles and gamma radiation is known as radioactivity. The decay of radioactive nuclei is given by $A(t) = A(0)e^{-kt}$.

Rate of chemical reactions

If two reactive chemicals are mixed to form a third chemical the rate of the reaction will depend on the concentrations of reactants and product. Chemical reactions can be zero order, first order and second order.

Radioactivity

The unstable nuclei of certain chemical elements emit energy in the form of radioactive decay with the emission of alpha and beta particles and gamma radiation. The basic law of such decay is that the rate is directly proportional to the amount of chemical present. If we let $A(t)$ represent the amount of radioactive material present at time t then from what has just been stated:

$$\frac{dA(t)}{dt} \propto A(t).$$

That is,

$$\frac{dA(t)}{dt} = -kA(t)$$

with the solution

$$A(t) = A(0)e^{-kt}.$$

Because the radioactive material is decaying, its amount is decreasing and so we introduce the minus sign to ensure that the **decay constant**, k, is positive.

Different radioactive materials decay at different rates and in order to compare the vigor of one decay with another a comparative measure has been devised. This is known as the **half-life** of the material (denoted by the lower case Greek character τ) and it is the time it takes for an amount to decay by 50%. Once we know the half-life of a radioactive material then we can find the specific value of for that material. That is, if

$$A(\tau) = A(0)/2$$

then

$$A(\tau) = A(0)/2 = A(0)e^{-k\tau}.$$

Hence, $e^{-k\tau} = 0.5$ and so

$$-k\tau = \ln\left(\frac{1}{2}\right) = -\ln 2$$

giving

$$k = \frac{\ln 2}{\tau}.$$

In the life sciences, one application of radioactive decay is in the technique of **carbon dating**. All living matter contains two isotopes of carbon, namely ^{12}C and ^{14}C and the ratio of the amounts of one to the other remains a constant in any living organism. Now, ^{12}C is stable but ^{14}C is radioactive and so when the organism dies the amount of ^{12}C remains but the amount of ^{14}C decreases. Consequently, by measuring the change in the ratio of the amounts of ^{12}C to ^{14}C in a piece of dead organic matter we can deduce the elapsed time since it died. This is the principle behind the technique.

Another application is in radon mitigation. Radon is a radioactive gas formed by the natural decay of radium-226. It occurs particularly in areas underlain by granite and is considered to be a health hazard to homes in such locations. However, radon has a use in radiotherapy.

Exercises and answers

Exercise 1
In 1940 the remains of a fire were discovered in a cave near the French town of Lascaux. It was found that only 15% of the original amount of ^{14}C was present. Given that for ^{14}C the half-life is 5 570 years find how long ago the fire was lit?

Exercise 2
The half-life of the gas radon is 3.825 days. How long will it take for an amount of the gas to reduce by 75%?

Answer 1
Use $A(t) = A(0)e^{-kt}$, where $A(t) = 0.15A(0)$, so that $e^{-kt} = 0.15$. Now $k = \frac{\ln 2}{\tau}$, where $\tau = 5\,570$ years, so that

$$k = \frac{\ln 2}{5570} = 0.000124 \text{ year}^{-1}.$$

Therefore, since $e^{-0.000124t} = 0.15$ we see that

$$t = \frac{1.8971}{0.000124} = 15\,299 \text{ years.}$$

Answer 2
Use $A(t) = A(0)e^{-kt}$ where $A(t) = 0.25A(0)$ so that $e^{-kt} = 0.25$. Now $k = \frac{\ln 2}{\tau}$, where $\tau = 3.825$ days, so that

$$k = \frac{\ln 2}{3.825} = 0.1812 \text{ day}^{-1}.$$

Therefore, since $e^{-0.1812t} = 0.25$ we see that

$$t = \frac{1.3863}{0.1812} = 7.7 \text{ days.}$$

Chemical reactions

If two reactive chemicals (called reactants) are mixed to form a third chemical they will do so over a given period of time. The rate at which this reaction occurs is measured in moles per liter per second (mol l^{-1} s^{-1}) – the rate at which the concentration of the third chemical increases. In general, this rate will not be a constant but will vary in time. The rates at which the concentrations of each of the individual component reactants decreases, however, will not necessarily be the same as the rate at which the concentration of the resultant chemical increases – it all depends upon the ratio of the amounts of the individual reactants. For example, If two reactant chemicals A and B with concentrations $A(t)$ and $B(t)$, respectively, are mixed in the ratio 1:2 to form the chemical C with concentration $C(t)$ then we write

$$A + 2B \to C.$$

We now assume that the reaction is such that as the resultant chemical C is being created both component reactants A and B are simultanteously being depleted. If the rate of decrease of concentration of A at time t is $r(t)$ then at any instant

$$\frac{dA(t)}{dt} = -r(t)$$

(negative to indicate a decrease). However, because of the mix, there is twice the amount of B present so the rate of decrease of concentration of B twice as fast. That is

$$\frac{dB(t)}{dt} = -2r(t).$$

Furthermore, the rate of increase of concentration of chemical C is

$$\frac{dC(t)}{dt} = 3r(t)$$

(positive to indicate an increase) because there is three times the amount of C as there is of A. In general, for the reaction $aA + bB \to cC$,

$$\frac{dA(t)}{dt} = -ar(t),$$

$$\frac{dB(t)}{dt} = -br(t)$$

$$\frac{dC(t)}{dt} = cr(t).$$

Here we have three linked differential equations whose solutions will depend upon the form of reaction rate $r(t)$.

Zeroth order reactions

The simplest assumption that can be made about the value of the reaction rate is that it is constant. This is called the **zeroth order reaction**. That is, $r(t) = k$. This gives rise to the three differential equations:

$$\frac{dA(t)}{dt} = -ar(t),$$

$$\frac{dB(t)}{dt} = -br(t),$$

$$\frac{dC(t)}{dt} = cr(t),$$

with respective solutions

$$A(t) = -akt + A(0),$$
$$B(t) = -bkt + B(0),$$
$$C(t) = ckt + C(0).$$

First-order reactions

The zeroth order reaction implies that when the reactant chemicals are fully depleted the reaction stops abruptly. This is not observed in practice and the more common assumption is that the reaction rate depends upon the amount of reactant present. This is called the **first-order reaction**. For example, in the reaction $A \to B$

$$\frac{dA(t)}{dt} = -r(t),$$
$$\frac{dB(t)}{dt} = r(t)$$

and

$$r(t) = kA(t).$$

This gives rise to the two differential equations:

$$\frac{dA(t)}{dt} = -kA(t) \text{ and } \frac{dB(t)}{dt} = kA(t).$$

The first of these equations has the solution

$$A(t) = A(0)e^{-kt}$$

so that the second equation becomes

$$\frac{dB(t)}{dt} = kA(0)e^{-kt}$$

with the solution $B(t) = -A(0)e^{-kt} +$ constant. Now $B(0) = -A(0) +$ constant $= 0$ and so constant $= A(0)$, therefore

$$B(t) = A(0)\left[1 - e^{-kt}\right].$$

Second-order reactions

In a **second-order reaction** the rate depends upon the square of the amount of reactant present. For example in the reaction $A \to B$

$$\frac{dA(t)}{dt} = -r(t),$$
$$\frac{dB(t)}{dt} = r(t)$$

and

$$r(t) = kA^2(t).$$

This gives rise to the two differential equations:

$$\frac{dA(t)}{dt} = -kA^2(t) \text{ and } \frac{dB(t)}{dt} = kA^2(t).$$

The solution of the first of these equations can be found by separating the variables and is

$$A(t) = \frac{A(0)}{1 + A(0)kt},$$

so that the second equation becomes

$$\frac{dB(t)}{dt} = k\left(\frac{A(0)}{1 + A(0)kt}\right)^2$$

with the solution

$$B(t) = -\left(\frac{A(0)}{1 + A(0)kt}\right) + \text{constant.}$$

Now, $B(0) = -A(0) + \text{constant} = 0$ and so constant $= A(0)$, therefore

$$B(t) = \frac{(A(0))^2 kt}{1 + A(0)kt}.$$

Exercises and answers

Exercise 1

Show that for a first-order reaction A → B, where the rate of reaction is $r(t) = kA(t)$,

$$B(t) = A(0)\left[1 - e^{-kt}\right].$$

Exercise 2

Show that for a second-order reaction A → B, where the rate of reaction is $r(t) = kA^2(t)$,

$$B(t) = \frac{kt}{A(0) + kt}.$$

Answer 1

In the reaction A → B

$$\frac{dA(t)}{dt} = -r(t) \text{ and } \frac{dB(t)}{dt} = r(t).$$

When $r(t) = kA(t)$ this gives rise to the two differential equations

$$\frac{dA(t)}{dt} = -kA(t) \text{ and } \frac{dB(t)}{dt} = kA(t).$$

The first of these equations has the solution

$$\frac{dA(t)}{dt} = -kA(t).$$

We separate the variables and integrate

$$\int \frac{dA(t)}{A(t)} = -k\int dt$$

to obtain $\ln(A(t)) = -kt + K$ where K is the integration constant. This can be rewritten as

$$A(t) = e^{-kt+K} = e^{-kt}e^{K}$$

where $A(0) = e^K$ so that

$$A(t) = A(0)e^{-kt}.$$

The second equation then becomes

$$\frac{dB(t)}{dt} = kA(0)e^{-kt}.$$

Separating the variables and integrating

$$\int dB(t) = kA(0)\int e^{-kt}dt$$

we find that $B(t) = -A(0)e^{-kt} + J$ where J is the integration constant. Clearly, $B(0) = -A(0) + J$ and $B(0) = 0$. Therefore $A(0) = J$ giving the final solution as

$$B(t) = -A(0)e^{-kt} + A(0) = A(0)\left[1 - e^{-kt}\right].$$

Answer 2

In a second-order reaction the rate depends upon the square of the amount of reactant present. In the reaction $A \rightarrow B$

$$\frac{dA(t)}{dt} = -r(t) \text{ and } \frac{dB(t)}{dt} = r(t)$$

where $r(t) = kA^2(t)$. This gives rise to the two differential equations:

$$\frac{dA(t)}{dt} = -kA^2(t) \text{ and } \frac{dB(t)}{dt} = kA^2(t).$$

Considering the first equation:

$$\frac{dA(t)}{dt} = -kA^2(t)$$

then separating the variables yields

$$\frac{dA(t)}{A^2(t)} = -kdt$$

and so, integrating,

$$\int \frac{dA(t)}{A^2(t)} = -k\int dt$$

gives $-A^{-1}(t) = -kt + K$ where K is the integration constant. Now, $-A^{-1}(0) = K$ and so $-A^{-1}(t) = -kt - A^{-1}(0)$. That is

$$\frac{1}{A(t)} = kt + \frac{1}{A(0)}.$$

Rearrangement of this equation gives

$$A(t) = \frac{A(0)}{1 + A(0)kt}.$$

The second equation then becomes

$$\frac{dB(t)}{dt} = k\left(\frac{A(0)}{1 + A(0)kt}\right)^2.$$

Separating the variables and integrating:

$$\int dB(t) = k\int\left(\frac{A(0)}{1 + A(0)kt}\right)^2 dt$$

produces the result

$$B(t) = -\left(\frac{A(0)}{1 + A(0)kt}\right) + \text{constant.}$$

Now, $B(0) = -A(0) + \text{constant} = 0$ and so constant $= A(0)$, therefore

$$B(t) = -\left(\frac{A(0)}{1 + A(0)kt}\right) + A(0) = \frac{A(0)(1 + A(0)kt) - A(0)}{1 + A(0)kt} = \frac{(A(0))^2 kt}{1 + A(0)kt}.$$

J1 WHY USE STATISTICS?

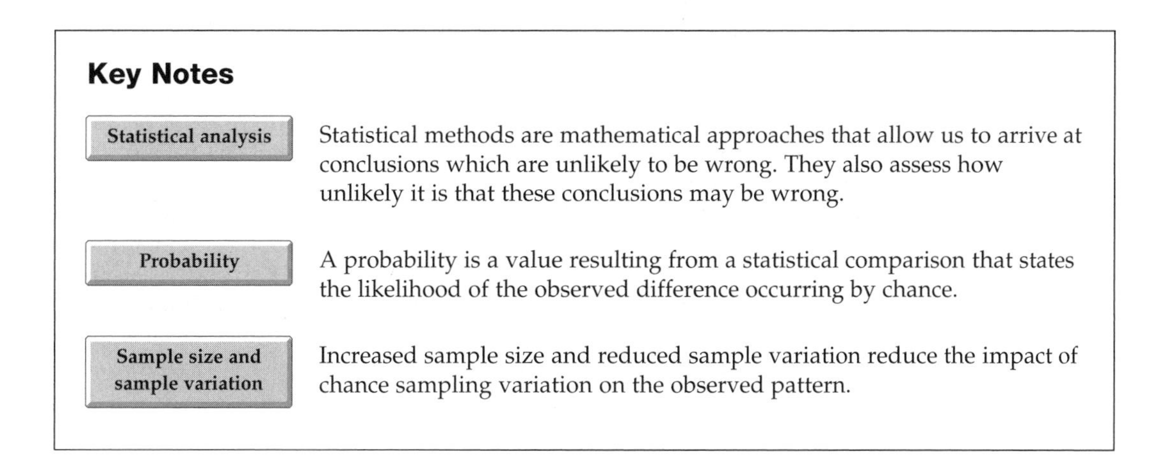

Key Notes

Statistical analysis	Statistical methods are mathematical approaches that allow us to arrive at conclusions which are unlikely to be wrong. They also assess how unlikely it is that these conclusions may be wrong.
Probability	A probability is a value resulting from a statistical comparison that states the likelihood of the observed difference occurring by chance.
Sample size and sample variation	Increased sample size and reduced sample variation reduce the impact of chance sampling variation on the observed pattern.

Statistical analysis

Statistical analysis is an essential tool in the scientific process. Almost all measurements are subject to some variation, and often this variation may prevent us from making unequivocal conclusions in comparisons between different measurements. Statistical methods are mathematical approaches that allow us to arrive at conclusions which are unlikely to be wrong. They also assess just how unlikely it is that these conclusions may be wrong.

Statistical analysis should be used whenever a **quantified assessment** of the relationship between sets of measurements is desired. Statistics need not be applied if

- single data points are to be compared (e.g.'Bob is shorter than Sam'),
- measurement error is several orders of magnitude less than the differences and the likelihood of erroneous conclusions is obviously negligible (e.g. 'adult domestic cats weigh less than adult lions'),
- the data displays no intra-class variation; for example, comparison of one unvarying protein with another.

Probability

A **probability, *p*,** is a value resulting from a statistical comparison that states the **likelihood of the observed difference occurring by chance**. For example a study may show that sun-adapted leaves have a higher stomatal density than shade leaves, with a probability of 0.01 ($p = 0.01$). (Recall that stomata are the holes in the leaf surface through which gaseous exchange occurs.) This means that, given the observed variation in stomatal density in sun-adapted and shade-adapted leaves, the calculated chance that these observations are part of the same population is 0.01, or 1% or 1 in 100 (these are just different ways of expressing '$p = 0.01$'). Given that 1 in 100 is fairly unlikely, we could be fairly confident in this case that the different was real. If the probability was $p = 0.001$ (or 1 in 1 000), then we could be even more confident.

On the other hand a probability of 0.1 indicates that the observed difference could, by chance, occur one time in ten, which is not sufficiently rare to be confident that the difference is real. Therefore, there is a **probability threshold** (often set at 0.05) for **statistical significance**, above which comparisons are said to be **nonsignificant**.

The term '**statistically significant**' is very important in science. Often, this phrase is reduced to the word 'significant', as in 'male babies are significantly heavier at birth than females'. The statement that a difference is statistically significant really is saying this: it is unlikely that this difference could have occurred by chance.

Statistical significance is said to occur at or below a threshold probability value, often called the **alpha-** or **α-level**. The alpha level is commonly set at **0.05** (or 5%). This value is arbitary, but is widely accepted. In some areas, for example clinical studies, where the risk of false conclusions may be dangerous or costly, more robust criteria may be employed, for example 0.01 or even lower. Of especial relevance in the days before statistical software could calculate probabilities, **critical values** are the values of a statistic that relate to a given probability. Thus, for a chi-squared test of a two-by-two table (see p. 165) the 5% (or 0.05) critical value is 3.841. All values of χ^2 at or above this value thus have a probability of less than 5%.

To summarize:

- Probabilities always lie between 0 and 1
- Probabilities less than or equal to the alpha-level (usually 0.05) are 'significant'.
- Probabilities greater than the alpha-level (usually 0.05) are non-significant.
- The smaller the probability, the more confident we can be that the observed pattern is not due to chance sampling variation.

To understand the role of **chance** consider the following two examples.

Firstly, if you take a set of ten random numbers between 0 and 100 and compare them to another set of ten random numbers also between 0 and 100, the usual result would be to find that there were no statistically significant differences between the sets (i.e. neither set was higher or lower), which is no surprise, given that both sets come from the same population. However, if this test is repeated again and again, it would be found that every so often the result would be significant: in fact, on average, one time in 20, the probability would be below 0.05. **Even when there is no real pattern in the data, chance variation will cause patterns to occur which appear to be statistically significant.** However, the chance of very highly significant probabilities occurring is unlikely, so a value of $p = 0.001$ will only occur by chance one time in a thousand.

Secondly, if we consider a statement such as 'the success in mating for male black grouse was positively correlated ($p = 0.05$) to the size of their territory', then the simplistic conclusion is easy: large territories are good for black grouse as they increase reproductive output. However, there is a possibility that we are reaching a mistaken conclusion and we know exactly what that possibility is: 0.05 or 1 in 20. What this means is that given the variation in the relationship between the size of the territory and the success in mating for black grouse, we would expect a positive correlation to arise one time in 20 even when the underlying population showed no pattern.

Sample size and sampling variation

Probabilities express the likelihood of the observed pattern in the data (say that potatoes grown in field A are larger than those grown in field B) could have occurred by chance sampling variation. Increased significance (lower probability values) will occur with: (1) increased sample sizes and (2) reduced sample variation. This is because both increased sample size and reduced sample variation reduce the impact of chance sampling variation on the observed pattern.

Thus, if the potato weights are very variable in both fields, then the variation among a sample of ten potatoes will be high too, and this will result in a weaker probability (i.e. closer to 1) whereas if the potato weights show limited variability within a field, there will be a smaller probability that field A potatoes are larger than field B potatoes. Equally, if a sample size of ten is chosen, then a few anomalous specimens will distort the mean weight of the sample, but if a sample size of 100 is used, the sample mean weight will more closely match that of the mean of the whole field and is less likely to be distorted.

The consequence of the impact of sample size and sample variability on significance levels is that it will be easier to detect patterns in your data if the sample sizes are large and the sample variability is low. Reducing the underlying sample variation and choosing the right sample sizes are critical components of **experimental design**.

J2 Experimental design

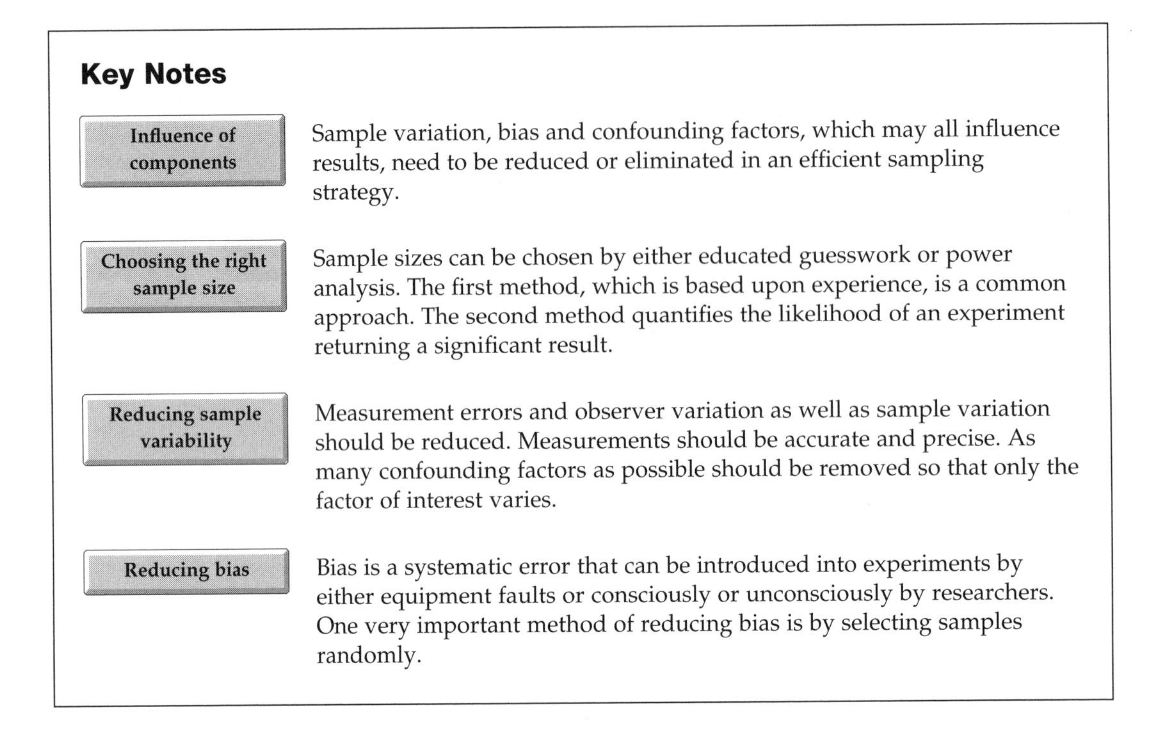

Key Notes

Influence of components
Sample variation, bias and confounding factors, which may all influence results, need to be reduced or eliminated in an efficient sampling strategy.

Choosing the right sample size
Sample sizes can be chosen by either educated guesswork or power analysis. The first method, which is based upon experience, is a common approach. The second method quantifies the likelihood of an experiment returning a significant result.

Reducing sample variability
Measurement errors and observer variation as well as sample variation should be reduced. Measurements should be accurate and precise. As many confounding factors as possible should be removed so that only the factor of interest varies.

Reducing bias
Bias is a systematic error that can be introduced into experiments by either equipment faults or consciously or unconsciously by researchers. One very important method of reducing bias is by selecting samples randomly.

Influence of components

Experimental design is the process of most efficiently designing an experimental set-up (or the sampling strategy for an observational study) to extract the patterns we are interested in. To discover the patterns of interest, we need to reduce or eliminate the influence of three components: sample variation; bias and confounding factors. Sample variation arises from two components: sample size (large sample sizes will have lower variation) and the underlying sample variation.

Choosing the right sample size

It is clear that large samples are better than small ones, but it is usually costly to have very large sample sizes, and therefore usually we have to compromise. There are two ways to choose sample sizes: **educated guesswork** and **power analysis**.

Educated guesswork sounds a rather unscientific way to proceed for something so important, but it is in fact the usual approach in many areas of life sciences. Experienced researchers will usually design experiments based on their experience. By finding a similar type of experiment in the literature, you can assess the general area in which your sample size should be. However novel you imagine your work to be, you can always find some published work which will give you a useful indication.

Power analysis is the quantification of the likelihood of an experiment returning a significant result, usually expressed as a percentage. Thus, an experimental design with a power of 80% would result in a significant result 80 times out of a 100 were the experiment to be repeated. The power depends on sample size, sample variability, significance level (commonly 0.05) and '**effect size**'. The effect size is the magnitude of the impact of a factor on the variable. For example, we might anticipate that a particular diet will result in mice that are 10% heavier, and we can then calculate the power of an experiment of a given sample size under this assumption. Power analysis for simple experimental designs are readily available in computer programs or from formulas in more advanced statistical texts. To apply these, it is necessary to either run a pilot experiment or to consult the literature, to obtain estimates of sample variation and effect size. Just as with choosing the significance level, deciding what power is acceptable is a somewhat arbitrary exercise, but a target of above 85% can be achieved.

This discussion assumes that there is no confusion over what is meant by sample size. Usually, there is no problem with this, but one possible pitfall is **pseudoreplication,** which occurs when separate measurements or counts are inherently linked in a way which means they are **not fully independent**. For example, if we had two tanks each containing ten developing tadpoles but fed on different diets, we could record the size of each tadpole at five different time points, say on days 0, 5, 10, 15 and 20. This would total 100 measurements ($5 \times 10 \times 2$) but they would not be 100 independent measurements: each tadpole is being measured repeatedly, so a tadpole that was large at day 0 might well also be large at all the other time points. Further, there might be a tank effect other than the diet, but as all tadpoles on one diet are grown in the same tank, it would not be possible to discern this. Ideally, pseudoreplication should be avoided, and experiments should be designed with this in mind, sometimes it is not possible.

Reducing sample variability

Reducing sample variability involves two components: reducing measurement error and reducing the underlying sample variation. Reducing measurement error requires that attention is paid to ensuring that the tools employed in measurement are **accurate** and **precise** (see Topic B1) and that observer variation (the variation caused by different individuals collecting data) is minimized.

Reducing the underlying sample variability involves removing as many confounding factors as possible, so that only the factor of interest varies. Confounding factors are commonplace in all biological systems. It is very common for a set of variables to vary in unison, such as high sunlight levels and high temperature. If you are interested in the effects of sunlight levels independently of temperature, then it would be necessary to design an experiment (say in a temperature-controlled greenhouse) which removed the effects of temperature. The heart of every well-designed experiment should be protocols which remove or attempt to remove the effects of any factor which is likely to influence the variable of interest. Discerning which factors may confound the data needs careful thought and, sometimes, good biological insights. Not only should the physical environment be carefully regulated, but also the status (e.g. age, sex, size, genotype) of the experimental subjects which may influence the variable.

In **observational studies** reducing the underlying sample variability must be differently managed because, unlike experiments, no manipulation of the subjects occurs, and this can be viewed as a weakness of observations compared to experiments. However, carefully designed sampling strategies can help statistical detection of underlying patterns. Two approaches are, firstly **stratified**

sampling, where sample variation is reduced by only selecting a subset of the population, e.g. plant populations on south-facing slopes, or female adults between the ages of 20 and 25 years; and secondly, **co-factor measurement**, which is the measurement of any factors (e.g. age, sex, slope, temperature, water pH) which are not the key subject of the study, but which may influence the outcome.

Sophisticated statistical analysis may be able to partition the variation in the underlying data due to these confounding factors and allow detection of the variation due to the factor of interest in the study. For example, a survey on the impact of the number of previous births to one mother on birth weight in humans may wish to allow for race and maternal weight as potential factors affecting birth weight. In the first instance, before employing the more complex statistical tests, simple tests may be employed to detect the influence of the suspected confounding factors. If these can be seen to exert no or minimal influence, then the analysis can be concluded simply.

Reducing bias

Bias is a systematic error, such as overestimation of the weight of a sample. It can be introduced by faulty measuring equipment, or by conscious or unconscious bias on the part of researchers or human subjects.

Randomization

One very important tool in avoiding bias is **randomization**, which is the random selection of experimental samples, to ensure their **independence**. If we cannot be sure that samples have not been selected independently and without bias, our faith in the outcome of any analysis should be shaken.

Imagine we wish to consider the impact of introducing a new soya bean variety on the fungal damage the crop may suffer. We could grow both varieties in an environmentally controlled glasshouse to limit the impact of variation in temperature and humidity, expose the plants to fungal spores and finally record the plant dry weight over 4 months. Even in a carefully managed situation like an environmentally controlled glasshouse, there will be many factors that could affect the plant's growth rate and which have nothing to do with the fungus. For example, the greenhouse may be brighter and warmer at one side than the other; the air-conditioning outlet may be closer to some plants than others, producing patches of cool air; and the automatic watering regime may be uneven in its operation. If we arranged the pots of beans such that those with the old soya bean variety were together at one side of the glasshouse whilst those with the new variety were all at the other, the differences due to the varietal resistance to fungus would be confounded by the patchiness of the glasshouse environment. If, however, plants were randomly allocated on a grid to ensure no systematic bias, then we could have much greater confidence in the results. Other examples of non-random sampling are shown in *Table 1*.

How to randomize your experiment

To ensure your experimental design is randomized, you need to locate some random numbers and apply those to your sampling program. Random numbers are easy to locate. They can be found either from printed tables, a scientific calculator, a general spreadsheet program (such as Microsoft® Excel) or from a specialized statistics software packages. Printed tables are tedious and usually unnecessary in the modern age. Calculators have the advantage of being simple and transportable, but can be tedious in all but the most modest of experiments.

Table 1. Examples of non-random sampling

Sample population	Sampling method	Possible biases
Class of 100 students	The first ten to enter the classroom	The keenest students? Those best able to read timetable? Those not possessing disability? Those not attending class in previous session?
Tank of 50 goldfish	First 10 to be caught by net	Slowest, least alert? Possibly ill, dying or congenitally stupid? Alternatively, more active fish in upper water may be caught, which may be sex or aged biased.
Daisy plants in a lawn	Nearest to haphazardly thrown marker ball.	Plants more likely to be in hollows? Away from trees? All similar distance from starting point (e.g. path) and possibly suffering from similar disturbance levels?
Grid of bean plants in plant pots in a glasshouse	Nearest row of pots	More similar light levels, watering regimes, pesticide exposure, etc. than in whole population.
Adult males in a city	Questionnaire on sexual behavior sent to random individuals, but only 30% are returned	More boastful may respond? Those embarrassed may not?

Using a spreadsheet to generate a table of exactly the right number of random numbers is simple and convenient, and allows you to print out this table and insert it into you laboratory or field notebook. See below for some tips on **generating random numbers**.

Approaches to randomization

There are two simple approaches to randomization. Firstly, If you have **a smallish population** from which to sample (for example, 100 students), then a good approach is to allocate a number to each individual. Then you can select a random individual by generating a random number in the range 1 … N (in this case, 100), and this can be repeated until you have reached your sample size. If you generate the same random number again, that individual will have already been selected. Just ignore repetitions and generate another random number.

Secondly, if you have a **large population** or a **population of undefined size**, such as the population of snails in a large field, an alternative approach is to randomly choose the location of the sampling site. In this case you will need to generate two numbers, to specify the x- and y- coordinates of a corner or the center of the sampling unit, e.g. a square quadrat. This may necessitate laying out tape measures to delineate the area, or a global positioning system (GPS) device may be used to locate the sites.

Blind and double-blind methods

In the case of experiments involving human subjects, it may be desirable to avoid the risk of unconscious bias influencing the outcome by using a **blind** design. This is where the subject does not know whether they are in the experimental or control group. For example, in a drug trial, all individuals are given a pill, but the control group are given a **placebo** which has no drug content. This reduces the chance of individuals reporting an effect due to psychosomatic autosuggestion. A further level of sophistication is the **double-blind** design, in which the researcher assessing the subject is unaware of whether that individual is in the experimental or control group, and therefore is not swayed in their assessment by that knowledge.

Generating random numbers

Many scientific calculators and spreadsheet programs have a function which generates random numbers, usually in the range 0 to 1, which may be labeled 'RAN#' or 'RND' or 'RAND' or similar (check what the function really does on your software or calculator by quick trial-and-error).

To **generate a random number in the range 0 to** N, two steps must be taken:

(1) Multiply the random number in the range 0 to 1 by N. This will produce a random number in the range 0 to N.
(2) Round the number to the nearest whole figure.

Alternatively, to **generate a random number in the range 1 to** N, we need to add a subtle tweak to this:

(1) Multiply the random number in the range 0 to 1 by $N - 1$. This will produce a random number in the range 0 to N – 1.
(2) Add 1 to this figure. This will now be transformed to a random number in the range 1 to N.
(3) Round the number to the nearest whole figure.

In Micosoft Excel, these steps can be achieved in one line. The function RAND() returns a random number between 0 and 1 and the function ROUND([number], *d*) rounds [number] to *d* decimal places. To round to a whole number, we want zero decimal places, so the equation to type in a Excel spreadsheet cell is:

= ROUND(RAND()*23,0)

to get whole random numbers between 0 and 23, or

= ROUND(RAND()*50,0)

to get whole random numbers between 0 and 50, or

= ROUND((RAND()*99)+1,0)

to get whole random numbers between 1 and 100. To produce a set of such numbers, simply paste this equation into further spreadsheet cells.

J3 TESTS AND TESTING

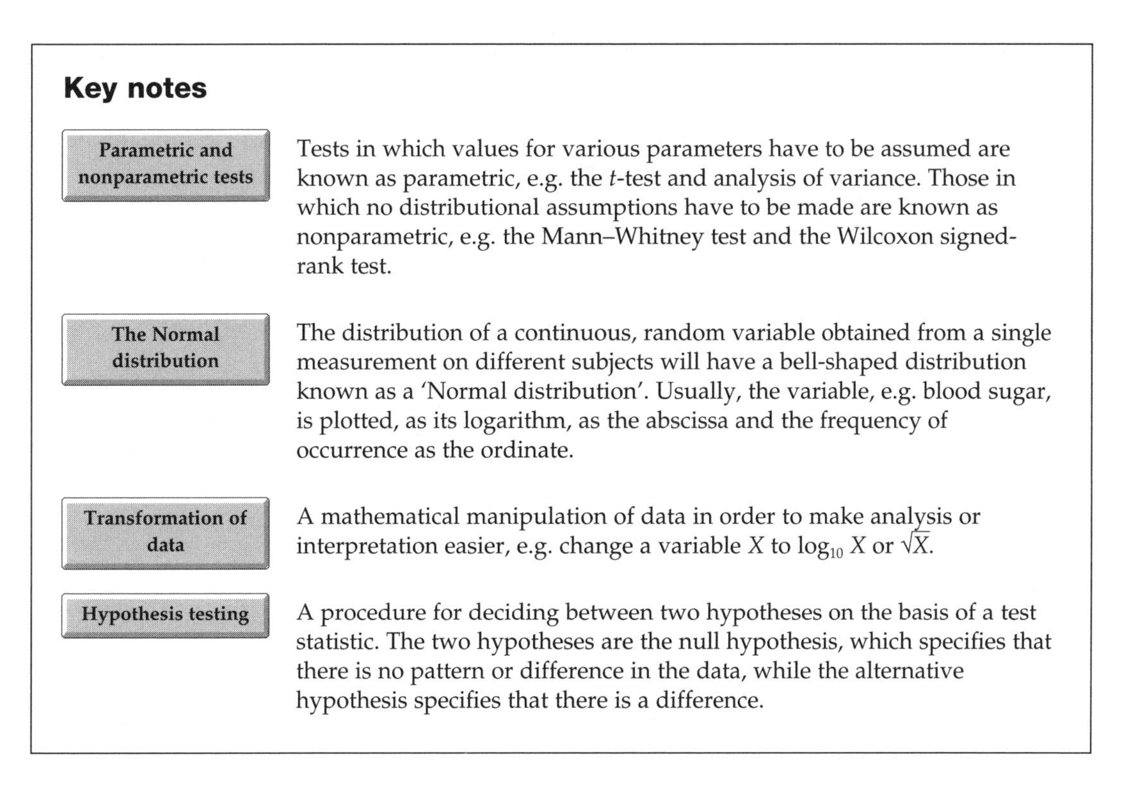

Key notes

Parametric and nonparametric tests

Tests in which values for various parameters have to be assumed are known as parametric, e.g. the *t*-test and analysis of variance. Those in which no distributional assumptions have to be made are known as nonparametric, e.g. the Mann–Whitney test and the Wilcoxon signed-rank test.

The Normal distribution

The distribution of a continuous, random variable obtained from a single measurement on different subjects will have a bell-shaped distribution known as a 'Normal distribution'. Usually, the variable, e.g. blood sugar, is plotted, as its logarithm, as the abscissa and the frequency of occurrence as the ordinate.

Transformation of data

A mathematical manipulation of data in order to make analysis or interpretation easier, e.g. change a variable X to $\log_{10} X$ or $\sqrt{X}$.

Hypothesis testing

A procedure for deciding between two hypotheses on the basis of a test statistic. The two hypotheses are the null hypothesis, which specifies that there is no pattern or difference in the data, while the alternative hypothesis specifies that there is a difference.

Parametric and nonparametric tests

Statistical tests can be divided into two types: **parametric** and **nonparametric**. Parametric tests are more powerful but make more demanding assumptions about the data; in particular they require (1) the distribution of errors in the data to be **normal** and (2) the data to be **continuous** (see Topic B1)**.** Examples of parametric tests are the *t*-test and analysis of variance because they assume that the data is approximately **normally distributed** (i.e. a bell-shaped curve). If the distribution of data points deviates from normality, you should attempt to **transform** the data (see p. 145).

Tests that make no distributional assumptions about a population under investigation are known as **nonparametric**. Such tests are usually very simple to perform. If your data is **ordinal** (i.e. see Topic B1) or is extremely non-normal, even after transformation, you should employ a **nonparametric test.**

The Normal distribution

Many formal statistical tests reiterate a mantra that you must always check for normality before employing parametric tests. That such tests (such as Kolmogorov–Smirnov, Anderson–Darling or Shapiro–Wilks tests) are not routinely available in all software packages (though the more comprehensive usually do possess one of these), and that the results are almost never cited in the literature may make you skeptical as to how rigorously these tenets are kept to. Furthermore, there are some difficulties in applying tests: small data sets are less likely to show significant deviation from a normal distribution because of sample size limitations (but more likely to be biased by a single outlier) and large data sets are quite unlikely not to, as the slightest discrepancy will give rise to a significant deviation.

A question which arises is: is normality necessary? One answer is no. Many parametric statistics, particularly the analysis of variance, are robust to deviations from normality.

The ***F*-ratio** which is the summary statistic of the analysis of variance can be viewed as a **noise-to-signal ratio**. This *F*-ratio is entirely independent of the underlying distribution; it is merely the probability which assumes normality. If the *F*-ratio is 20, this means the model (say that adult height depends on gender) accounts for 20 times as much variation in the data as the remaining 'error' or 'noise'. If you have an *F*-ratio of 20 (or higher) you can be fairly confident in the conclusions of your analysis, even if the associated probability may be slightly erroneous.

Nevertheless, you should be aware that gross deviations from normality may cause your analysis to be flawed, particularly if different samples or populations possess distributions markedly different from each other. Further, others may expect you to check for normality.

Testing if your data meets normality assumptions

Step 1

In order of robustness, either

(1) Test that your data is not significantly deviant from a normal distribution using a test designed do this (such as the Kolmogorov–Smirnov, Anderson–Darling or Shapiro–Wilks tests);
or
(2) Plot the data on a Normal probability plot. Perfectly normal data will lie on a straight line. Almost no real biological data will do this, but you only need to be concerned about gross deviations;
or
(3) Construct a frequency histogram of your data to see whether it looks normal, i.e. symmetrical and monomodal (i.e. single-humped).

Step 2

If the data is not deviant from normality according to the test in Step 1, you can now progress to using parametric statistics.

If the data is deviant from normality you have to choose between the following.

(1) Using a nonparametric test. However, there may not be an appropriate test readily available. Also, nonparametric tests are less sensitive than parametric. So it may be better to use a parametric test if you can.
(2) Transforming the data. Transformation is the process of applying a mathematical conversion to each data point, such that it conforms (better) to a normal distribution. *Table 1* lists types of transformation.
(3) Use a test based on another distribution. If you are using a more sophisticated statistical package then you may be able to fit your data another distribution (e.g. Weibull). This is fairly advanced ground, and beyond this text.

Table 1. Types of transformation

Transformation	Type of data applied to
log Y or log$(Y+1)$	Contagious or clumped distribution, or where factors in an analysis of variance are synergistic and apparently multiplicative. (Use +1 if data contains zeros or numbers near zero.)
arcsin $\sqrt{Y}$	Percentage (0–100%) or proportional (0–1) data. (Note that arcsin often appears on calculators as $\sin^{-1}$.)
$\sqrt{Y}$ or $\sqrt{Y+0.375}$	Distributions where the variances are proportional to the means, i.e. they increase in unison. (Use the +0.375 form if data contains zeros or numbers near zero.)

(4) Use a parametric test anyway. If the data will not transform satisfactorily, and there is no available suitable nonparametric test, you may decide to press ahead with a parametric test anyway. You should recognize that the test may be inaccurate, and you should be wary of accepting as significant results very close to $p = 0.05$. You may wish to increase the alpha-level (e.g. to 0.01) to compensate. The analysis of variance is a particularly robust test and will not be easily distorted by non-normality.

Transformation of data

Data transformation is the mathematical manipulation of the data in order to make it easier to analyze or interpret. Commonly, transformation aims to either (1) **linearize** (to make nonlinear patterns linear (see Topic E2)); or (2) **normalize** (to make non-normal distributions better approximate normal distributions); and (3) **equalize variances** (to make data conform to an assumption of the analysis of variance that sample variances are equal (or 'homoscedatic').

There are many different possible transformations to normalize and linearize but most are rarely applied. The commonest are listed in *Table 1*.

Hypothesis testing

The formal description of the process of performing a statistical test invokes **hypothesis testing**. There are always two competing hypotheses in any situation: the **null hypothesis**, often labeled H_o, and the **alternative** or **active hypothesis**, H_a.

The null hypothesis is that there is no pattern or difference in the data, whilst the active or alternative hypothesis is that there is a difference. Examples of null and alternative/active hypotheses are given in *Table 2*.

Table 2. Examples of the null and alternative/active hypotheses

Comparison	Null hypothesis	Alternative/active hypothesis
Weights of two different strains of mice when fed under standard conditions.	Individuals of the two strains have *the same weight.*	Individuals of the two strains have *a higher (or lower) weight.*
Density of field buttercup in relation to soil moisture levels.	Field buttercup density is *unaffected* by soil moisture level.	Field buttercup density is *related to* soil moisture level.
A sprinter's performance in relation to training regimes A and B.	A sprinter's performance is *unaffected* by the training regime.	A sprinter's performance is *higher (or lower)* under training regime B.

When you obtain a probability from you statistical test which is significant (i.e. at or below the **alpha-level**, commonly 0.05), then the formal statement which follows is that you are able to **'reject the null hypothesis'**. If, in contrast, the probability is above the alpha-level, the result is nonsignificant and you are **'unable to reject the null hypothesis'**. Note that a nonsignificant result does not indicate that that the null hypothesis should be accepted. This is an important point: there are many factors (e.g. poor sampling, inefficient experimental design, sample size too small) that can lead to an insignificant result.

Whilst formal hypothesis descriptions are commonly used in training scientists, they are not so apparent in the scientific literature, where it is more usual and useful to discuss results as demonstrating (or not demonstrating) significant differences or patterns, and null hypotheses are rarely mentioned.

Two kinds of mistake: Type I and Type II errors

When you have tested a null hypothesis, you will either accept or reject it, as outlined above. There are two kinds of mistaken conclusion that the test may cause you to make: either

(1) to detect a significant pattern when there is no real effect – a Type I error (i.e. incorrectly reject the null hypothesis), or
(2) fail to detect a pattern when one does exist – a Type II error (i.e. incorrectly accept the null hypothesis).

The chance of there being a Type I error is specified by the threshold probability level, α, usually set at 0.05 (see Topic K1). For a given sample size, the more robust the attempts to avoid a Type I error (for example by reducing the value of α) the more likely it is that a Type II error is made. The usual value of $\alpha = 0.05$ is thus a reasonable compromise.

Presenting statistical information

When presenting the results of a statistical analysis, it is important to communicate the information in a way that is precise and succinct. There are three pieces of information which should always be present when stating a statistical result: (1) the statistic, (2) the degrees of freedom and (3) the probability. What do we mean by these terms?

The **statistic** is the calculated value (such as t in a t-test, or F in an analysis of variance, or r in a correlation or regression, or U in a Mann–Whitney U-test).

The **degrees of freedom** (variously labelled df, dof or υ) is the sample size n minus the number of constraints. Commonly dof $= n - 1$ or $n - 2$. F-statistics from an analysis of variance have a pair of dof values. The degrees of freedom may be written as e.g. 'dof = 34' or within subscripted brackets, e.g. $t_{[34]} = x$ (where x is some value).

We have discussed **probability** in Topic J1. However, when expressing this information, it is usual to only give the precise probability of the result that is significant, otherwise it is sufficient to state 'ns' to indicate nonsignificant. With significant results, you can either express probabilities as exact numbers (e.g. $p = 0.031$), or to classify them into groups of increasing significance: for example: $p < 0.05$, $p < 0.01$, $p < 0.001$. (Note that these descriptions are incomplete, for when we say $p < 0.05$ we mean $p < 0.05$ and $p > 0.01$.) When dealing with a large number of statistical tests at once, it is convenient to use this classification approach, but as modern statistical software allows calculation of exact probabilities, it is usually better to use these exact figures.

Examples of presentation of statistical results are given in *Table 3*.

Table 3. Examples of the presentation of statistical results

Statistical test	Example
t-test	$t = 3.223$, dof = 23, $p = 0.002$
Analysis of variance	$F_{[1,46]} = 15.369$, $p < 0.0001$
Correlation (Pearson's)	$R_{[13]} = -0.706$, $p = 0.0033$
Chi-squared	$\chi^2_{[1]} = 5.522$, $p = 0.016$

Interpreting computer software output

Many students come to statistics lectures and understand (more or less!) what is said to them there, then panic when they try to interpret the output from statistical software. This may be because the output contains reams of supplementary information, or because the format is not familiar. If you can locate the three components discussed above (value of the statistic, degrees of freedom and probability), then you have the key information you need.

K1 SEARCHING FOR PATTERNS AND CAUSES

Key Notes

Correlation	A correlation assesses the association between two variables. The correlation coefficient *r* expresses the degree of association, and lies between –1 and 1.
Linear regression	Linear regression assesses how closely a set of data fits a straight line. It is the relationship between a dependent variable and one independent variable, and is sometimes described as 'calculating the line of best fit'. The degree of fit is expressed by the coefficient of determination, r^2.
Multiple regression	Multiple regression describes the relationship between the dependent variable and two or more independent variables.
Nonlinear regression	Data that is better fitted by using a curve are known as nonlinear. Curves may be described by various functions, such as exponentials, logarithms, polynomials and powers.
Correlation or regression?	Never calculate both a correlation and a regression from the same set of data. One of these must be inappropriate to your data set.

Correlation

A **correlation** assesses the **association** between two variables. The **correlation coefficient *r*** expresses the degree of association, and lies between –1 and 1. A value of –1 indicates that there is a perfect negative association between variables; a value of 0 indicates that there is no association between variables; a value of +1 indicates that there is a perfect positive association between variables. The relationship is shown in *Table 1*.

A **correlation** assesses the **association** between two variables (such as leaf size and herbivore damage). It **does not presume that one variable depends on the other**. A significant correlation **does not indicate causation**, that is to say, because we do not know whether variable 1 affects variable 2 or vice versa (or indeed, that both operate via a third variable), we cannot deduce the cause of an observed pattern.

Which correlation coefficient?

There are two types of correlation commonly used: the **Pearson's** or **product–moment correlation** (a **parametric** test) (*r*) and the **Spearman's rank correlation (**r_s**)** (a **nonparametric** test). Choosing which is appropriate depends on the usual criteria when deciding on whether the data meets the assumptions of a parametric test (see p. 143). You should use Spearman's rank correlation if the data does not come from a (roughly) **normal** population, or if the data is

Table 1. The meaning of the correlation coefficient

Value of correlation coefficient, *r*	Meaning	Graphical illustration
–1	There is a perfect negative association between the variables (i.e. as one gets larger the other gets smaller, in perfect proportion)	
0	There is no association between the variables	
1	There is a perfect positive association between the variables (i.e. as one gets larger the other gets larger, in perfect proportion)	

ranked or **ordinal** (see Topic B1). If the data meets the assumptions of normality, you should use Pearson's test, as this will be more sensitive.

As an example consider the sex ratios and intersexual weight differences in bees and wasps (the *Hymenoptera*). In some species, females are markedly heavier than males. Amongst the same species, it is not uncommon to find that males outnumber females. Can we find a significant association when comparing the weight ratio (the weight of males relative to females) with the sex ratio (defined as the proportion of males in the population)? *Table 2* shows data from eight different bee and wasp species. Plotting this (*Fig. 1*) shows a pattern. Inputting this data into an analysis package gives a Pearson's correlation coefficient of –0.813, with 6 degrees of freedom and a probability of 0.014, which we can express as $r_{[6]} = -0.813$, $p = 0.014$. This indicates that there is a significant negative association between sex ratio and weight ratio in these eight species of bee and wasp.

Linear regression

Linear regression assesses how closely a set of data fit to a straight line, and is sometimes described as 'calculating the line of best fit'. The degree of fit is expressed by the coefficient of determination, r^2. This is the proportion of the

Table 2. Relationship between weight ratio and sex ratio for some species of bees and wasps

Weight ratio	Sex ratio
0.66	0.68
0.63	0.62
0.62	0.74
0.83	0.59
1.25	0.50
0.65	0.63
0.66	0.61
0.75	0.65

(Data are from Trivers and Hare. *Science* 1976; **191**:249–263.)

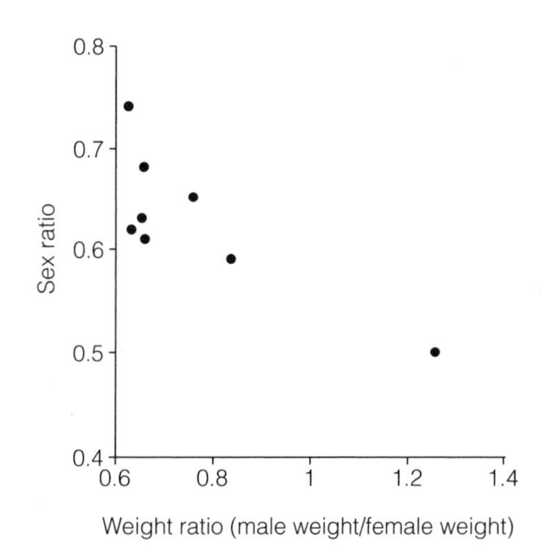

Fig. 1. Relationship between sex ratio and weight ratio for bees and wasps.

total variation in the dependent variable y that the independent variable x explains, and varies between 0 and 1 (it is often expressed as a percentage, between 0% and 100%). The line of best fit is calculated in the form $y = mx + c$ (see Topic E2). The overall **significance** of the regression can be summarized in an F statistic. However, we may be interested in asking whether (1) the **slope** (m) is significantly different from zero or (2) whether the **intercept** (c) is significantly different from zero, and these can be tested using a t-test (see p. 157). It is usual for statistical software to perform all these tests automatically. It may not be necessary to your study to cite all these. (See 'Interpreting results' in the following example.)

A regression is applied to a pair of variables, when we can regard one variable as independent. If an experiment is carried out in which one variable is controlled and another variable is measured, then the controlled variable is independent. An example would be if plant growth was measured in controlled-environment cabinets at different temperatures (e.g. 5°C, 10°C, 15°C and 20°C).

As an example consider the weight loss in flour beetles at different humidity levels. Groups of 25 beetles were kept without food in containers at fixed controlled humidity levels, and their weight loss over a 6 day period was recorded. The results are given in *Table 3*. It is easy enough to enter our two columns of data into our analysis package and to select the regression option, but make sure that you are clear which variable is the dependent (y) and which the independent (x). Remember the independent variable should be the one which was controlled at a series of different values during the experiment. In this case the relative humidity is the independent variable. This is not just a matter of plotting the figure the right way round; it is a statistical mistake.

Table 3. Relationship between weight loss and humidity for flour beetles

Relative humidity (%)	Weight loss (mg)
0	8.98
12	8.14
29	6.67
43	6.08
53	5.90
62	5.83
75	4.68
85	4.20
93	3.72

(Data are from Sokal and Rolhf. *Biometry*. Freeman, NY, 1981.)

Interpreting results

Table 4 shows a typical regression analysis table from a statistical software package (after analysing the data above). It may look intimidating, but once you've grasped the key points, it should be less so! The software you use may give this information in a slightly different format, and may contain some other information and/or omit some, but if you can find your way around this example, you should be able to decode them too.

Summarizing the analysis

For the equation of the line, when you have located the values of m and c (see ❺) you can simply insert these into the equation format of $y = mx + c$ to obtain $y = -0.053x + 8.698$. The coefficient of determination, r^2, is 97.8%. The overall statistical analysis, as stated in ❾ is $F_{[1,7]} = 266.1$, $p = 0.000\,01$. Now you have a complete summary the regression:

$$Y = -0.053x + 8.698,\ r^2 = 97.8\%,\ F_{[1,7]} = 266.1,\ p = 0.000\,01.$$

On occasions you may wish to state whether the slope or intercept deviate significantly from zero, in which case you cite the appropriate t-statistic and the associated probability. Finally, you may wish to plot the data and the regression line, as shown in *Fig. 2*.

Multiple regression

A detailed coverage of multiple regression is beyond the scope of this text, but although the mathematical complexity can become quite involved, the basic principle is easy to understand.

A basic linear regression describes the relationship between the dependent variable (the thing we are interested in) and **one independent variable.** In contrast, **multiple regression** describes the relationship between the dependent variable and **two or more independent variables.** For example, the algal density in a lake may depend on a variety of environmental factors, such as temperature, nitrate levels and phosphate levels. Each of these parameters can be treated as an independent variable in a multiple regression.

Table 4. A typical regression analysis table

❶ DEP VAR:WTLOSS ❷ N: 9 ❸ MULTIPLE R: 0.987 SQUARED MULTIPLE R: 0.974
ADJUSTED SQUARED MULTIPLE R: 0.971

❹ Variable	❺ Coefficient	❻ Standard error	❼ *T*	❽ *p* (two-tail)
Constant	8.697 578	0.191 594	45E + 02	0.000 00
Relative humidity	−0.053 270	0.003 265	−16E + 02	0.000 00

❾ Analysis of variance

Source	Sum of squares	Degrees of freedom	Mean square	*F*-ratio	*p*
Regression	23.512 171	1	23.512 171	266.153 534	0.000 001
Residual	0.618 384	7	0.0883 41		

❶ Indicates the dependent variable is 'weight loss'. Always try to give your variables meaningful names.
❷ Indicates the sample size.
❸ The following three components are *r*, the coefficient of determination r^2, and an 'adjusted' coefficient of determination, $r^2_{[adj]}$. Cite r^2 unless you are undertaking a multiple regression (see below), when you should always cite $r^2_{[adj]}$.
❹ This table contains (1) the components (*m* and *c*) of the line of best fit and (2) the analyses to test whether these components differ significantly from zero. The first column is labeled 'variable' and distinguishes the 'constant' (the *c* or intercept of the line) and the slope (*m*) relating to the independent variable, which here is the relative humidity.
❺ This gives the coefficients of *c* (= 8.697...) and *m* (= −0.053...).
❻ The standard error of the intercept *c* and the slope *m*.
❼ These are the *t*-test statistics, to determine whether the values of *c* and *m* are significantly different from zero.
❽ These are the related probabilities, indicating both are highly significantly different from zero. Note that although the software has run out of decimal places, there will be a non-zero digit at the end of these numbers. If you cite probabilities like this, it should be in the form $p < 0.000\,01$, not $p = 0$.
❾ This table is an analysis of variance (see Topic K2) of the overall regression. An ANOVAR table like this is explained on p. 157, but all you really need to focus on is the probability (*p*). When citing the result, you should state the *F*-ratio, the degrees of freedom (1 and 7) and the probability, like this : $F_{[1,7]} = 266.1$, $p = 0.000\,01$

A general **multiple regression equation** takes the form

$$Y = b_1x_1 + b_2x_2 + b_3x_3 + \ldots + c,$$

where x_1, x_2, x_3 ... are the independent variables and b_1, b_2, b_3, ... the associated coefficients. The coefficient of each independent variable tells us what relation that variable has with *y*, the dependent variable, with all the other independent variables held constant. So, if b_1 is high and positive, that means that if x_2, x_3 etc. do not change, then increases in x_1 will correspond to large increases in *y*.

As with a simple linear regression, we can calculate a coefficient of determination r^2, for a multiple regression. However, in the case of a multiple regression, this value is confounded by the number of independent variables, such that if the number of variables increases, the value of r^2 inevitably rises, until when there are as many independent variables as data points, r^2 will always be 100%.

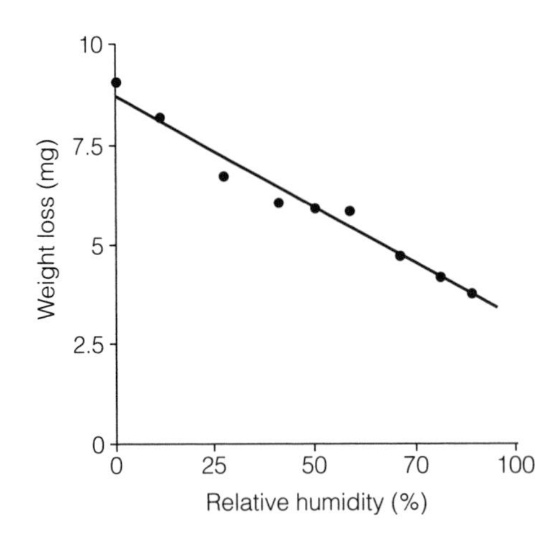

Fig. 2. The relationship between weight loss and environmental humidity in groups of 25 flour beetles.

This problem is overcome by using an **'adjusted' coefficient of determination, $r^2_{[adj]}$** which removes this effect. Be wary of any multiple regression which cites r^2 and not $r^2_{[adj]}$!

When constructing a multiple regression, it is common to have a list of possible independent variables to use in the regression model. Some of these may have little predictive value and offer nothing useful to the model. To determine which variables to keep, a procedure called **stepwise regression** is adopted. There are two different approaches: either to start with all the variables in the model and sequentially remove those with least influence on the dependent variable ('backward stepwise') or to start with one independent variable and sequentially add the most predictive of the 'spare' variables to the model.

Nonlinear regression

Just as we may wish to identify the line of best fit to a straight line (linear regression), so we may want to determine the line of best fit to a curve. Software makes this computationally demanding exercise simple to accomplish. Packages commonly offer a range of different function types, such as exponential, logarithmic, polynomial and power (for a full description of these terms, and examples, see Topic E2). The degree of fit can be assessed by the coefficient of determination, r^2.

Which function should you use?

If you suspect that the data may better fit a nonlinear function than a straight line (for example because a scatter plot seems to be curved), you may wish to consider applying a curved function. If so, you should bear in mind the following points.

Think about the biological meaning of the function; for example, a cubic polynomial equation ($y = a + bx + cx^2 + dx^3$) will provide a (slightly) higher r^2 for the beetle weight loss data above than the linear fit, but it is hard to imagine what biological meaning is added. If you have reason to expect a particular function to be appropriate (for example, an exponential decay in enzyme activity over time), then you should apply the equation this knowledge suggests.

Choose the most parsimonious solution. If you can achieve almost as good a fit with a straight line, or with a two-parameter curve as opposed to a three-parameter curve, then choose the simpler solution.

Be aware that, as with the multiple regression (above), adding parameters to an equation will inevitably increase the r^2 value. Some packages may calculate a 'corrected' r^2 which may compensate for this effect.

Should you transform data to make it linear?

Prior to the wide availability of computing power, it was standard practice to transform data to make linear regression possible. For example, data approximately fitting a logarithmic function could be logarithmically transformed. Is this a preferable approach?

If you wish to understand the relationship between the dependent and independent variables, it may be easier to interpret untransformed data which has a nonlinear line fitted. However, there are circumstances when it is useful and/or conventional to present transformed plots; for example, it is easier to compare different straight lines than different curves (see Topic E1 for examples) and it is conventional to transform enzyme data in the Michaelis–Menten relationship (see Topic E2).

Correlation or regression?

If you want to find the relationship between sets of data that you can express as a scatter plot, as in *Fig. 3*, then a correlation or regression (not both) will be appropriate. You should never calculate both a regression and a correlation from the same set of data: one of these must be inappropriate to your data set.

A **correlation** assesses the **association** between two variables (such as leaf size and herbivore damage). It **does not presume that one variable depends on the other**. For example, do larger leaves attract more herbivore activity? Do nasty-tasting leaves deter herbivores and grow bigger? We don't know. A **regression** presumes that one variable (or factor) is **independent**, and that the other variable did not affect it. If you **controlled** one variable and measured another, then the controlled variable is truly **independent** of the other and a regression is the correct approach.

Examples of data appropriate for correlation and regression are given in *Table 5*.

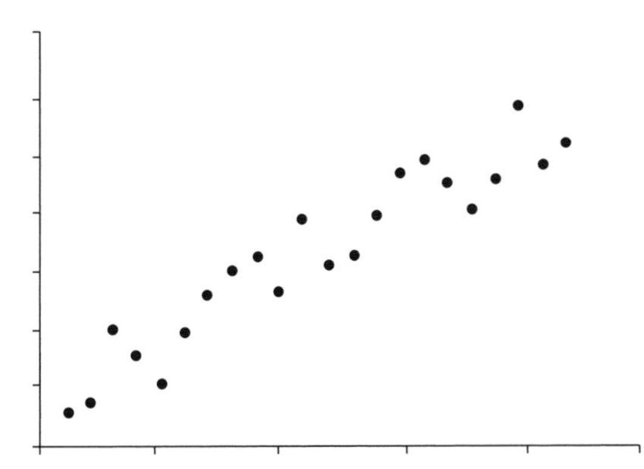

Fig. 3. A typical scatter plot that would be suitable for either correlation or regression analysis.

Table 5. Examples of data that are appropriate for correlation and regression

Regression	Correlation
Controlled nitrogen levels and plant growth	Observed soil nitrogen levels and plant growth
Controlled insecticide levels and insect mortality	Insect density and insectivorous bird density
Controlled temperature levels and cell development rate	Blood hemoglobin levels and sporting performance
Set time intervals and drug response	Observed time of first flowering and number of flowers produced

K2 LOOKING FOR PATTERNS IN CONTINUOUS DATA

Key Notes

Analysis of variance

Analysis of variance (ANOVA) tests whether normally distributed data from two or more independent samples are likely to be drawn from populations with the same mean value. The main output from ANOVA is the probability of obtaining the data you have if the samples are drawn from populations with the same mean.

Multiple comparisons

If you are comparing the mean values of three or more samples of data, you will want to make more than one comparison. For example if you have three samples (A, B and C) you will want to compare A with B, B with C and C with A. This increases the risk of making a false claim. To avoid this problem we reduce the probability of making a false claim (the significance level) on each test. The simplest (but not the best) way to do this is to divide the significance level by the number of comparisons being made.

The Mann–Whitney–Wilcoxon test

The Mann–Whitney–Wilcoxon test is used to investigate whether two independent samples of data are likely to have been drawn from populations with the same median value. The median is less sensitive than the mean to outliers, or skewness in the distribution. This test is therefore useful if you have doubts about the normality of your data.

Analysis of variance

The first small-sample test for normally distributed data was created by William Gosset in 1908. Gosset was working at the Guinness brewery in Dublin when he devised the test. Guinness refused to allow him to publish his test, but he published anyway under the name 'A Student'. The test has since become known as 'Student's *t*-test'. Student's test only works for two samples and cannot be extended to three or more.

In 1925, Ronald Fisher, who worked at Rothamsted agricultural research station (near Harpenden, Hertfordshire) extended Gosset's test to more than two samples and to the analysis of more sophisticated experimental designs. Fisher called his method analysis of variance. When analysis of variance is used on data from two samples, it gives exactly the same results as Student's *t*-test. Gosset's test for two independent samples is therefore just an historical curiosity now, but you will still find it being used by people who enjoy historical curiosities.

Thus, if you have data from two or more independent samples and you have reason to believe that the results are drawn from a normally distributed population, then you can use analysis of variance to test whether the two samples are drawn from populations with the same mean value. Fisher devised a variety of experimental designs that were primarily intended to show which crop varieties gave the greatest yield. ANOVA was the analytical technique he created to handle the results returning from the experiments.

ANOVA operates by looking at all the variability in a data set and attributing the variation either to differences between the different crops or to random variation. What it tests is whether the results obtained from an experiment are consistent with the hypothesis that all the crop varieties have the same mean yield. ANOVA can, of course, be used on any other type of experimental data with equally good results.

One-way analysis of variance

Table 1 shows a subset of data from an experiment to determine the mean butterfat content (%) of milk from several breeds of cattle. Looking at the figures suggests that the Ayrshire gives lower butterfat than the Jersey, or the Guernsey, but there seems to be some overlap between the Jersey and Guernsey.

Table 1. Mean butterfat content (%) for various breeds of cattle

Breed	Butterfat content (%)				Mean
	Sample 1	Sample 2	Sample 3	Sample 4	
Ayrshire	4.44	4.37	4.25	3.71	4.1925
Guernsey	5.30	4.50	4.59	5.04	4.8575
Jersey	5.75	5.14	5.25	4.76	5.2250
		Grand mean			4.7583

(Data are from Sokal and Rohlf. *Biometry*. Freeman, New York, 1995.)

The way in which the variability in the data is estimated is to calculate the overall mean of all the measurements (the grand mean), then to look at each measurement in turn and see by how much it differs from the grand mean. These differences are then squared and added up to give a total sum of squares. (There are good theoretical reasons why we square the differences, but the simple reason is that it makes them all positive.)

The **total sum of squares** is

$$\begin{aligned}&(4.44-4.7583)^2+(5.30-4.7583)^2+(5.75-4.7583)^2\\&\quad+(4.37-4.7583)^2+(4.50-4.7583)^2+(5.14-4.7583)^2\\&\quad+(4.25-4.7583)^2+(4.59-4.7583)^2+(5.25-4.7583)^2\\&\quad+(3.71-4.7583)^2+(5.04-4.7583)^2+(4.76-4.7583)^2\\&\quad=3.448.\end{aligned}$$

This is a measure of the total amount of variability in the data. If we want to measure the random variation in the experiment, then we should compare each measurement, not with the grand mean, but with the mean value for that particular breed of cattle. This will give us the **random variability**, or **error sum of squares**, which is

$$\begin{aligned}&(4.44-4.1925)^2+(5.30-4.8575)^2+(5.75-5.2250)^2\\&\quad+(4.37-4.1925)^2+(4.50-4.8575)^2+(5.14-5.2250)^2\\&\quad+(4.25-4.1925)^2+(4.59-4.8575)^2+(5.25-5.2250)^2\\&\quad+(3.71-4.1925)^2+(5.04-4.8575)^2+(4.76-5.2250)^2\\&\quad=1.257.\end{aligned}$$

To estimate the variability between the breeds, we could pretend that each measurement gave the mean yield for that breed of cattle. This would eliminate any error variation and leave us with the variation between breeds.

The between-breeds sum of squares is

$$
\begin{aligned}
&(4.1925 - 4.7583)^2 + (4.8575 - 4.7583)^2 + (5.2250 - 4.7583)^2 \\
&\quad + (4.1925 - 4.7583)^2 + (4.8575 - 4.7583)^2 + (5.2250 - 4.7583)^2 \\
&\quad + (4.1925 - 4.7583)^2 + (4.8575 - 4.7583)^2 + (5.2250 - 4.7583)^2 \\
&\quad + (4.1925 - 4.7583)^2 + (4.8575 - 4.7583)^2 + (5.2250 - 4.7583)^2 \\
&\quad = 2.191.
\end{aligned}
$$

Drawing these together gives:

between breeds sum of squares = 2.191
error sum of squares = 1.257
total sum of squares = 3.448

The total sum of squares is the sum of the between-breeds sum of squares and the error sum of squares. This is not coincidental. We have successfully attributed the variability to either variation between the butterfat content of the three breeds, or to the random variation between the individual animals.

We are not there yet because the sum of squares is just that: it is a sum and as we include more measurements into the experiment, the sum of squares will get larger. What we want is an average, or 'mean square', that should stay the same. If you look at the table above for calculating the total sum of squares, you will see that there are 12 measurements in it. So it might seem reasonable to calculate a mean square by dividing the total sum of squares by 12. However, if you look closely you can see that this table also includes the grand mean. If you read down the 12 measurements, then the first 11 are unpredictable, but the 12th must be the number that causes the grand mean to be 4.7583. We say that this table has 11 degrees of freedom (df). So we can calculate the mean square by dividing the sum of squares by 11.

If this calculation of numbers of degrees of freedom sounds complicated, try this for a simple example. I tell you that I am thinking of two numbers which have a mean value of 6. When I say that one of my numbers is 4, then you can work out that the other number must be 8. My data set contains two numbers, but once I have told you the mean value, only one of the numbers is free; there is only one degree of freedom.

Applying the same logic to the error sum of squares table gives us 12 measurements, but three mean values, so has 9 degrees of freedom. Between breeds we have three different readings, and one grand mean, so 2 degrees of freedom. Results are summarized in *Table 2*. Note that the sums of squares add up to the total as do the degrees of freedom. The mean squares do not add up to anything useful, so I have left out the total.

Table 2. Statistical readings for between-breeds data

Source	Sum of squares	Degrees of freedom	Mean square
Between breeds	2.191	2	1.096
Error	1.257	9	0.1397
Total	3.448	11	

The next move is to compare the between-breeds mean square with the error mean square. This should give a measure of how large is the variation between breeds compared with the random variation between individual animals. The obvious thing to do is to calculate

$$F = \frac{\text{between-breeds mean square}}{\text{error mean square}} = \frac{1.096}{0.1397} = 7.844,$$

where F (F for Fisher) is going to be a measure of how big the effect we are looking for is in comparison with the experimental scatter. Larger values of F mean we are more likely to be looking at a real effect. There are tables of F-values telling us when we are 95% certain that the effect we are seeing is real and 99% certain. These days it is rare to need these tables because statistics packages calculate the significance level for us. *Table 3* shows the ANOVA table for the butterfat data, produced by a popular package. This tells us that the probability of obtaining these results if there is no difference in population mean yield between the breeds is only 0.011, or about 1 in 90. A much more likely explanation is that the differences between the breeds are real.

Table 3. ANOVA data for butterfat yield, produced by a popular package

	Sum of squares	Degrees of freedom	Mean square	*F*	Significance
Between breeds	2.191	2	1.096	7.844	0.011
Error	1.257	9	0.140		
Total	3.448	11			

Multiple comparisons

This is all very well, but it does not tell us which differences are 'real'. We can see from the original table (*Table 1*) that the Ayrshire has a low fat yield and the Jersey has a high fat yield, with the Guernsey in between. Have we shown that the Ayrshire gives significantly less than the Guernsey, or less than the Jersey, or what? The answer is that we have shown that at the $p = 0.05$ significance level, there is a difference in mean score in there somewhere. We must use other methods to decide where the difference lies and how big it is. The simple way to do this would be to take the breeds two at a time and look to see if the difference between them is significant. The results of doing this are shown in *Table 4*. This works, after a fashion, but the danger is that we are making three comparisons, and the risk of finding a significant difference that is not real is increasing with every comparison. If we had five breeds and the 10 possible comparisons, then this method would quite probably create a false significant result. Statisticians differ about the right way to correct for this. There are methods to suit all tastes. The simplest method is the **Bonferroni test**. All this requires you to do is to decide on a base significance level, say 0.05. You then count up the number of

Table 4. Differences in butterfat yields between different breeds of cattle

Difference	Mean yields	Difference	Significance, *p*
Jersey minus Ayrshire	5.2250 – 4.1925	1.032 5*	0.011
Guernsey minus Ayrshire	4.8575 – 4.1925	0.665 0	0.099
Jersey minus Guernsey	5.2250 – 4.8575	0.367 5	0.593

*Significant difference.

comparisons you are making (in this case, three) and divide the significance level by the number of comparisons.

$$\text{Bonferroni significance level} = \frac{\text{base significance level}}{\text{number of comparisons}} = \frac{0.05}{3} = 0.017$$

So in this case we would only accept a difference as significant if the p-value is less than 0.017. To a good approximation we are taking the same overall risk ($p = 0.05$) of falsely claiming to have found a significant difference. Looking back to the table of differences, we have one comparison with a significance level below $p = 0.017$; we have shown that milk from Jersey cows has a higher butterfat content than that from Ayrshire. The other differences are not significant.

Two-way analysis of variance

The great joy of ANOVA is that it can be extended to cope with data from more sophisticated experimental designs. For example, the data used above comes from cows that are only 2 years old. The experiment also included results from mature cows, as shown in *Table 5*. Looking at the mean butterfat yields for the different ages and for the three breeds gives the results shown in *Table 6*, which suggest that there is little difference between the 2 year old cows and the mature cows. The differences between the breeds seem to be about the same as before (*Table 7*).

Table 5. Comparison of butterfat (%) in mature and 2-year-old cattle from different breeds

Ayrshire		Guernsey		Jersey	
Mature	2 year	Mature	2 year	Mature	2 year
3.74	4.44	4.54	5.30	4.80	5.75
4.01	4.37	5.18	4.50	6.45	5.14
3.77	4.25	5.75	4.59	5.18	5.25
3.78	3.71	5.04	5.04	4.49	4.76

Table 6. Mean butterfat yields (%) for mature and 2-year-old cattle

Age	Mean	Number of samples, N
Mature	4.7275	12
2 years	4.7583	12
Total	4.7429	24

Table 7. Mean butterfat yield (%) for three breeds of cattle

Breed of cattle	Mean	Number of samples, N
Ayrshire	4.0088	8
Guernsey	4.9925	8
Jersey	5.2275	8
Total	4.7429	24

To pin this down we can use a two-way ANOVA which is calculated in exactly the same way as above, but with an additional calculation of the sum of squares between cows of different ages. The results (produced by a software package) are shown in *Table 8*. This table says that the total variability in the data is 11.375 units, of which over half (6.689) is caused by differences between the breeds and very little (0.006) by variation between the 2 year and the mature cattle. The variation between breeds is very highly significant; $p = 0.000$ (if we can believe the computer) while that between the 2 year and mature cattle is not significant ($p = 0.877$). We can therefore ignore the ages of the cattle as a factor and repeat our previous one-way ANOVA to look for differences between the breeds (*Table 9*).

So including the extra data has allowed us to show convincingly that the Ayrshire cows give milk with a lower butterfat content than either the Jersey or the Guernsey. We still have no reason to believe that there is a difference in butterfat yield between the Guernsey and the Jersey. If the ages of the cows had created a significant difference in butterfat yield, then we would have needed to consider both age and breed as factors. Analysis of variance methods exist to do this, but are beyond the scope of this text.

Table 8. Tests of between-subjects effects where the dependent variable is butterfat yield (%)

Source	Sum of squares	Degrees of freedom	Mean square	F	Significance
Age	0.006	1	0.006	0.024	0.877
Breed	6.689	2	3.344	14.292	0.000
Error	4.680	20	0.234		
Total	11.375	23			

Table 9. Results of one-way ANOVA to determine differences in butterfat yield between breeds of cattle

Difference	Mean yields	Difference	Significance, p
Jersey minus Ayrshire	5.227 5 – 4.008 8	1.218 7*	< 0.001
Guernsey minus Ayrshire	4.992 5 – 4.008 8	0.983 7*	0.001
Jersey minus Guernsey	5.2275 – 4.9925	0.235	0.993

The Mann–Whitney–Wilcoxon test

Where data are not normally distributed, we cannot use ANOVA. This is because ANOVA works by calculating the squares of the differences between individual measurements and the mean. A small number of outliers can have a disproportionate influence on the sums of squares. If we have, for example, a skewed distribution, then the results in the long tail of the distribution have a greater influence than we might wish to give them. We can allow for this type of problem by basing our tests not on the actual measurement, but rather on the position of the measurement relative to the other measurements we have taken – the **ranking** of the measurement.

In an experiment to assess the toxicity of a substance, ten experimental mice were injected with a low dose of the substance and the time (in seconds) for them to react to stimuli on their tails was measured. Ten controls that had been injected with a placebo were tested at the same time. The results are shown below in *Table 10*. The reaction times are not expected to be normally distributed and the extreme values confirm this expectation.

Rankings for the measurements from 1 to 20 are shown in the top row of *Table 11*. The second row in the table shows all 20 reaction times set down in order, from shortest to longest. There is a slight problem here because some of the measurements are the same. Where this has happened it is not possible to allocate a unique ranking to a measurement. Instead a median ranking is given to all similar measurements. For example the 8th and 9th ranking measurements are both 2.8, so each is recorded as ranking 8.5th.

Table 11 provides a visual presentation of the results. The Cs are bunched together on the left, with the Es predominantly on the right. There is some overlap in the middle of the table. It looks as though the experimental treatment does indeed increase the 'average' response time.

To get a numerical handle on the differences we can calculate the Wilcoxon statistic, W. This is done by adding up the rankings of one group of measurements. To make this easy, I shall add up the rankings of the control measurements:

$$\begin{aligned}\text{sum of rankings of Cs} &= W\\ &= 1+4+4+4+4+7+8.5+11.5+11.5+11.5\\ &= 67.\end{aligned}$$

The more the Cs are bunched on the left of the table, the smaller this statistic will be. In the extreme case of all the control measurements being smaller than any of the experimental, then all the Cs will be on the left and all the Es on the right and the Wilcoxon statistic, W, will be

$$\begin{aligned}W &= 1+2+3+4+5+6+7+8+9+10\\ &= 55.\end{aligned}$$

Table 10. Time (seconds) for mice to react to a stimulus

Controls	Experimental
2.4	2.8
3.0	2.2
3.0	3.8
2.2	9.4
2.2	8.4
2.2	3.0
2.2	3.2
2.8	4.4
2.0	3.2
3.0	7.4

(Data are from E. Shirley. *Biometrics* 1997; **33**: 386–389.)

Table 11. Rankings for the measurements in the toxicity experiment

<table>
<tr><td>1</td><td>2</td><td>3</td><td>4</td><td>5</td><td>6</td><td>7</td><td>8</td><td>9</td><td>10</td><td>11</td><td>12</td><td>13</td><td>14</td><td>15</td><td>16</td><td>17</td><td>18</td><td>19</td><td>20</td></tr>
<tr><td>2.0</td><td>2.2</td><td>2.2</td><td>2.2</td><td>2.2</td><td>2.2</td><td>2.4</td><td>2.8</td><td>2.8</td><td>3.0</td><td>3.0</td><td>3.0</td><td>3.0</td><td>3.2</td><td>3.2</td><td>3.8</td><td>4.4</td><td>7.4</td><td>8.4</td><td>9.4</td></tr>
<tr><td>C</td><td>C</td><td>C</td><td>C</td><td>C</td><td>E</td><td>C</td><td>C</td><td>E</td><td>C</td><td>C</td><td>C</td><td>E</td><td>E</td><td>E</td><td>E</td><td>E</td><td>E</td><td>E</td><td>E</td></tr>
<tr><td>1</td><td colspan="5">4</td><td>7</td><td colspan="2">8.5</td><td colspan="4">11.5</td><td colspan="2">14.5</td><td>16</td><td>17</td><td>18</td><td>19</td><td>20</td></tr>
</table>

E, experimental; C, control.

If all the control measurements are larger than any of the experimental the Cs will be on the right and W will be 155. If the control and experimental measurements are all the same, then all the measurements will have a tied rank of 10.5 and W will be 105. The closer the Wilcoxon statistic is to 55, the more likely it is that there is a real difference between the experimental and control measurements.

Smaller W values mean we are more likely to be looking at a real difference between the experimental and control measurements. There are tables of W-values showing us when we can be 95% certain and 99% certain that the effect we are seeing is real. The normal way of calculating W and assessing its significance is to use a computer package. *Tables 12* to *14* show the output from a popular package.

This confirms our calculation of $W = 67$ and shows that the difference in median reaction time between the control and experimental conditions is very highly significant ($p = 0.004$). Mice injected with the active agent (experimental) have an estimated median reaction time of 3.5 s and this is almost certainly longer than the estimated median reaction time of 2.3 s for the control mice.

Note that for historical reasons we have two widely used calculations for the statistic used in the Mann–Whitney–Wilcoxon test. The Wilcoxon W statistic is calculated as above; the Mann–Whitney U statistic is calculated in essentially the same way, but with the minimum possible value (55 in the above example) being subtracted from it. So the Mann–Whitney U statistic for the above test would be $U = 67 - 55 = 12$. The significance level and interpretation remain exactly the same. Most statistics packages print out both statistics when carrying out the test.

Note also that the Mann–Whitney–Wilcoxon test only works for two independent groups. If there are three or more groups you will need to use the Kruskal–Wallis test. This test performs what amounts to analysis of variance on the rankings.

Table 12. Reaction time (seconds)

Treatment	N	Median
Control	10	2.300
Experimental	10	3.500
Total	20	3.000

Table 13. Ranking of the test results for reaction time

Treatment	N	Mean rank	Sum of ranks
Control	10	6.70	67.00
Experimental	10	14.30	143.00
Total	20		

Table 14. Test statistics where 'treatment' is the grouping variable

Statistic	Reaction time (s)
Wilcoxon W	67.000
Z	–2.908
Asymp. sig. (two-tailed)	0.004

K3 LOOKING FOR PATTERNS IN COUNT DATA

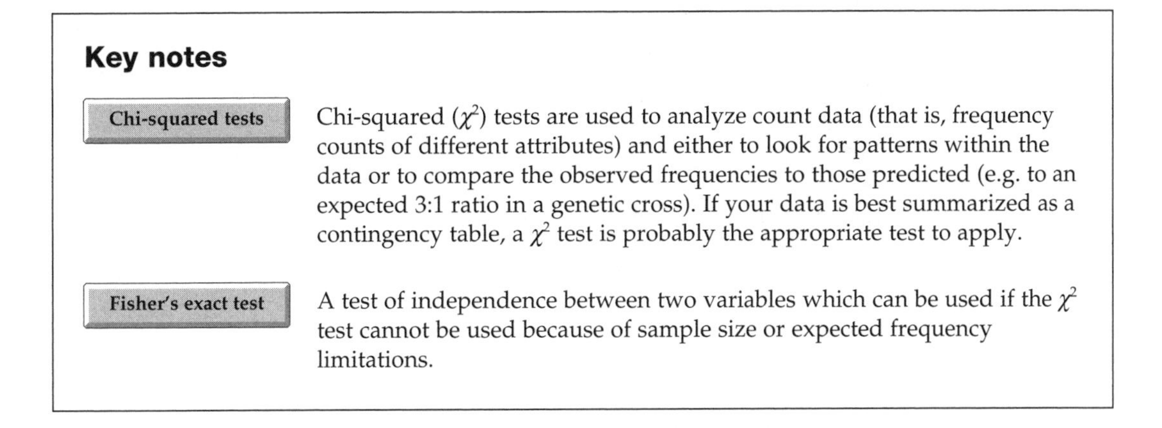

Key notes

Chi-squared tests

Chi-squared (χ^2) tests are used to analyze count data (that is, frequency counts of different attributes) and either to look for patterns within the data or to compare the observed frequencies to those predicted (e.g. to an expected 3:1 ratio in a genetic cross). If your data is best summarized as a contingency table, a χ^2 test is probably the appropriate test to apply.

Fisher's exact test

A test of independence between two variables which can be used if the χ^2 test cannot be used because of sample size or expected frequency limitations.

Chi-squared tests

It is appropriate to apply chi-squared analysis to a contingency table of frequencies of nominal or ordinal variables (see p. 3 for definitions), such as eye color, days of the week, winged or wingless, male or female. The test detects whether there is **heterogeneity** between the different classes. It is a **nonparametric** test, but makes the following assumptions: (1) the data was randomly sampled (or is a complete population); (2) the objects counted are independent of each other; and (3) the majority (say $>80\%$) of the **expected frequencies** should exceed 5

The third of these assumptions requires some expansion: in the course of calculating a chi-squared, the expected frequency for each cell is calculated. If you are using software to calculate χ^2, it should warn you if this restriction is broken. If your data has low or zero counts in some classes, **Fisher's exact test** can be employed.

The simplest data structure is a 2×2 contingency table such as *Table 1*. A 2×2 χ^2 statistic can be relatively easily calculated by hand or using a spreadsheet, unlike most of the other statistics treated here. *Table 2* shows a simplified approach that is easy to use and from which we can obtain a value for χ^2:

$$\chi^2 = \frac{n(ad - bc)^2}{(a+b)(c+d)(a+c)(b+d)}.$$

As a numerical example of a 2×2 chi-squared calculation consider the data given in *Table 3*. This will give a value for chi-squared of:

$$\chi^2 = \frac{n(ad - bc)^2}{(a+b)(c+d)(a+c)(b+d)}$$

$$= \frac{75(228 - 483)^2}{35 \times 40 \times 33 \times 42} = 2.5133.$$

Table 1. A simple 2×2 contingency table

	Blue eyes	Not blue eyes
Male	13	21
Female	22	36

Table 2. Simple approach to a 2×2 chi-squared statistic

	Blue eyes	Not blue eyes	Total
Male	*a*	*b*	*a* + *b*
Female	*c*	*d*	*c* + *d*
Total	*a* + *c*	*b* + *d*	*a* + *b* + *c* + *d*

Table 3. Example of a 2×2 chi-squared calculation

	Blue eyes	Not blue eyes	Total
Male	12	21	33
Female	23	19	42
Total	35	40	75

Table 4. Critical values of chi-squared for a 2×2 contingency table

Probability level, *p*	Critical value of χ^2
0.05	3.841
0.025	5.024
0.01	6.635
0.005	7.879
0.001	10.828

The value of the chi-squared statistic for this data is thus 2.5133. To check for statistical significance, we need to check in the table of critical values (*Table 4*). It can be seen that 2.5133 smaller than the critical value for $p = 0.05$, and therefore the observed pattern shows no significant heterogeneity: there is no effect of gender on the possession of blue versus non-blue eyes.

To use *Table 4* compare the calculated chi-squared statistic with the critical values. Remember, the higher the value of the calculated statistic, the lower (and hence more significant) the probability level. The probability level is given by the highest threshold the calculated value passes. If the value of the χ^2 is less than 3.841, the probability is greater than 0.05, so the result is nonsignificant. If the value of χ^2 is greater than 10.828, probability is less than 0.001 and should be stated as $p < 0.001$. If the value lies between 3.841 and 10.828, the probability level is given by the highest critical value exceeded. For example, if the calculated χ^2 is 6.721, this is above the 0.01 level of 6.635 and below the 0.005 level of 7.879, so $p < 0.01$.

Chi-squared tests can also be used to compare **distributions to a theoretical expectation**. For example, flower colors in a cross between two pink-flowered parent snapdragon (*Antirrhinum*) plants may be tested to see if they conform to simple Mendelian inheritance with red co-dominant with white, giving a 1:2:1 red:pink:pink:white expected ratio in the offspring (*Table 5*). The calculated χ^2 value is 2.891 (df = 3) is nonsignificant. The data does not significantly differ from the expected distribution.

Table 5. Chi-squared test for the distribution of flower color

	Red flowered	Pink flowered	White flowered
Observed	19	56	23
Expected	25%	50%	25%

Fisher's exact test

When a 2×2 chi-squared test cannot be applied due to sample size or expected frequency limitations, Fisher's exact test can be used. It is relatively easy to calculate the equation is given here by hand or with a spreadsheet. Alternatively, use an on-line calculator on the web (search for 'Fisher's exact calculator'). For example, using the contingency table summary as given in *Table 2*, the probability can be calculated directly by the following formula:

$$p = \frac{(a+b)!(c+d)!(a+c)!(b+d)!}{(a+b+c+d)!a!b!c!d!}.$$

K4 SEARCHING IN A DATA POND

Key Notes

Multivariate statistics

Multivariate statistics simultaneously analyze with several (or many) related variables. They are useful for detecting patterns in complex data; for example, the composition of plant communities or multiple metabolic parameters potentially associated with disease.

Types of multivariate statistics

There are two broad groups of techniques: (1) those that do not assume that there are dependent and independent variables, e.g. cluster analysis; and (2) those that do assume there are dependent and independent variables, e.g. multiple regression.

Multivariate statistics

Multivariate statistics are a range of techniques that **simultaneously** deal with **several related variables** and consider the inter-relationships between all those variables. These techniques are used to seek for patterns in complex data sets which are not easily assessed by more simple analyses. This section does not attempt to provide more than a brief insight into the range of multivariate approaches.

Different subtle definitions of the term 'multivariate' mean there is no clear agreement as to whether **multiple regression** (discussed on p. 152) and/or multivariate **analysis of variance** (discussed on p. 157) constitute 'proper' multivariate techniques. In so far as they are techniques which allow messy data to be thoroughly trawled, they belong to the family.

A fundamental component of many multivariate analyses is to **reduce the dimensionality** of the data. This would be useful, for example, in a study of diving rockhopper penguins, which collected 22 different parameters from each of 5 950 dives, and which required a 22-dimensional plot to display all variables at once. However, because many of the variables are inter-related, it is possible to summarize much of the variance in the data in only a few axes. The rockhopper penguins were sampled from three very different environments: subtropical (AMS), subantarctic coastal (KER) and subantarctic deep water (CRO), with widely differing food availability, food types and water temperature. The variables collected on each dive are shown in *Fig. 1*. A **principal components analysis** was applied to the data which reduced the data to a few 'principal components'. The first three of these encapsulated 65% of the total variance in the data. (see *Fig. 1*).

Types of multivariate techniques

Multivariate techniques can be divided into two broad groups.

(1) Techniques which reduce dimensionality but do not assume that there are dependent and independent variables, including **cluster analysis** (which classifies observations in groups), and a range of methods which can be

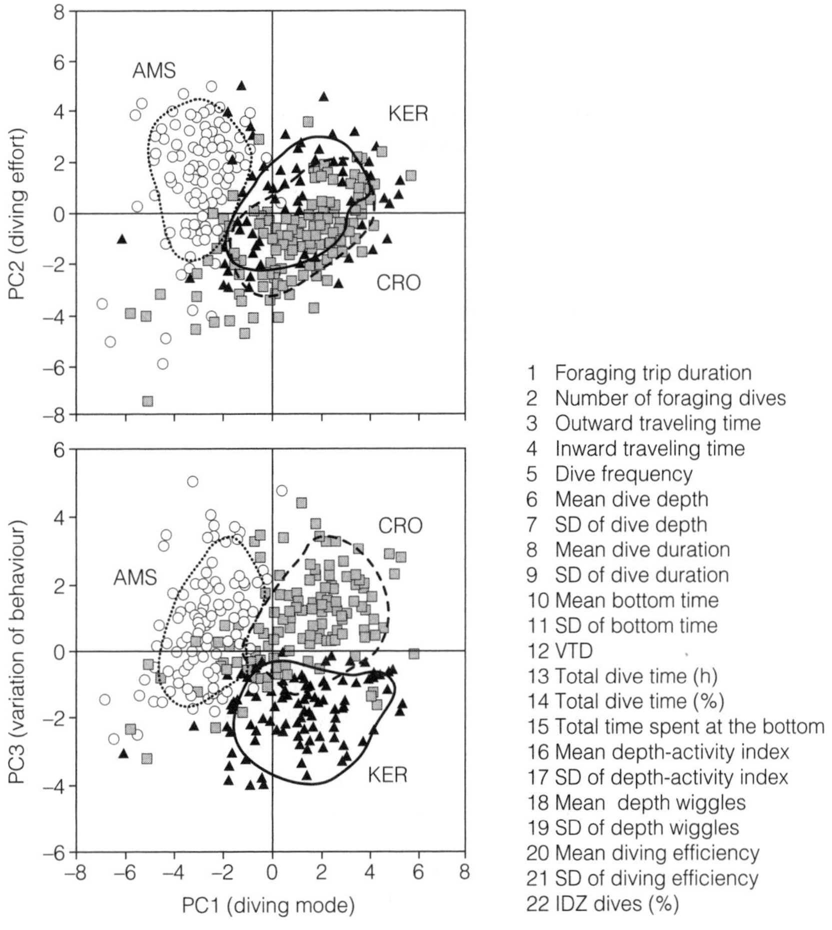

Fig. 1. Twenty-two parameters recorded in rockhopper penguins during dives in three different environments: subtropical (AMS), subantarctic deep water (CRO) and subantarctic coastal (KER). The first three axes of the principal component analysis summarize 65% of the total variation in the data and demonstrate that diving behavior differs markedly across the three environments. Each of the three axes can be broadly described as summarizing diving mode, diving effort or variation of behavior, but these are simplifications to aid interpretation.

grouped under the title of **ordination** or **scaling** techniques, including **principal components analysis, multidimensional scaling** and **correspondence analysis**. There are no hard-and-fast rules as to which of these ordination techniques should be applied in a given situation, although correspondence analysis is particularly suited to abundance data and hence is commonly used by plant community ecologists.

(2) Techniques which assume that there are dependent and independent variables, such as **multiple regression**, **multivariate analysis of variance**, **canonical correlation analysis** and **discriminant function analysis**. These methods are quite different from each other, and one could not be substituted for another.

Table 1 lists features of the various techniques.

As the underlying mathematical manipulations behind most of these techniques is complex, and because there may be a variety of different approaches

Table 1. Features of multivariate analysis techniques

Technique	Dependent variables		Independent variables (always multiple)	Use
	Number	Continuous or discontinuous	Continuous or discontinuous	
Multiple regression	Single	C	C and D	Quantifies the impacts of each of a suite of independent variables on the dependent variable.
Discriminant function analysis	Single	D	C	Quantifies the discrimination of separation of data into two or more **known** groups. It may be used to confirm the usefulness of groups found by ordination techniques.
Multivariate analysis of variance	Multiple	C	D	
Canonical correlation analysis	Multiple	C and D	C and D	Quantifies the relationship between two sets of variables.

C, continuous; D, discontinuous.

and choices that can be made within each technique, different software packages may give slightly (or greatly) differing results. More so than with simple statistical techniques, it is essential that the software package used is cited when multivariate results are presented. Some of these techniques are available in standard statistical packages such as SPSS, SAS, MINITAB and Systat, but there are also some packages dedicated to particular multivariate approaches, such as CANOCO (canonical correlation) DECORANA ('detrended' correspondence analysis). There are also a variety of esoteric programs which perform particular tasks according to fanciful rules (such as TWINSPAN, a vegetation data clustering program). Use these with care.

One particular application of cluster analysis and related techniques is in the construction of evolutionary phylogenies, either using molecular data or physical measurements (e.g. *Fig. 2*). There is a whole literature and software industry

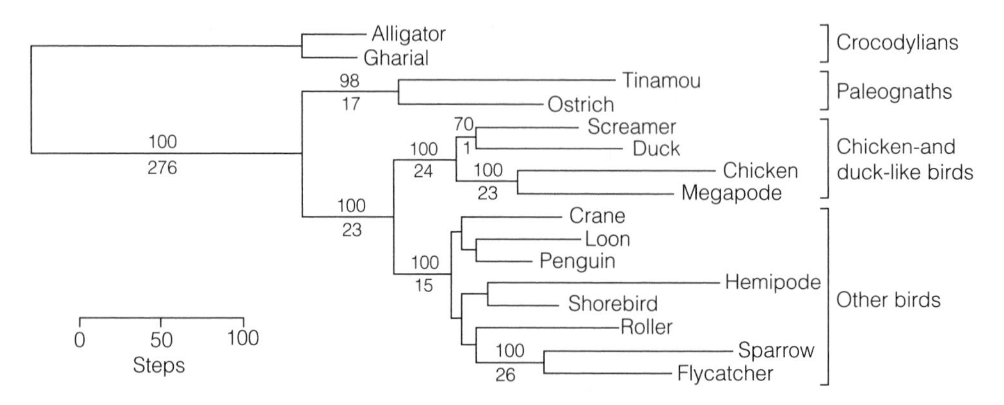

*Fig. 2. Divergence in the nuclear RAG-1 gene among basal groups of birds and crocodilians. (*From: Groth and Barrowclogh. *Molecular Phylogenetics and Evolution.* 1999; **12:**115–123.*)*

dedicated to considering the finer nuances of these approaches. It is worthwhile being aware that different tree-creation models can produce very different outcomes.

INDEX

absorbance of light 39, 40–1
acceleration 71, 72–5
accuracy 4–5, 20
acidity 39
active null hypothesis 145
adjusted coefficient of determination 154
algebra
  algebraic expressions 26–8
  difference equations 28
  differential equations 28
  inverse operations 27
  manipulating numbers 25–8
  operations 27
alkalinity 39
allometric scaling law 42–3
alpha levels 136, 146
alternative null hypothesis 145
analysis of variance (ANOVA) 157–60
  hypothesis testing 147
  multivariate 169–70
  two-way 161–2
angles, trigonometry 29–30
ANOVA *see* analysis of variance
area 9, 87–9
arithmetic sequences 106
attrition 113, 120–2

bar charts 21, 22
bases, logarithms 35
Beer's law 39, 40–1
bias 4, 138, 140–2
blind methods, bias reduction 142
Bonferroni tests 160–1
boundary values 96

calculators 16–17
calculus
  differential equations 95–105
  differentiation 55–64, 65–6
  integration and integrals 65–94
canonical correlation analysis 169, 170
carbon dating 128
carrying capacity, population growth 117
cause and pattern searches 149–56
Celsius scale 8
change of variables 76–8
chemical kinetics 127–33
  radioactivity 127–8
  rate of chemical reactions 127, 128–33
chemical reaction rates 127, 128–33
chi-squared tests 147, 165–7

cluster analysis 168–71
co-factor measurements 140
collating data 15, 20–1
collecting data 15
colorimetry 41
common difference 106
common ratio 106
component influence 137
compound units 6–7
concentration
  Beer's law 41
  measurements 10–14
  SI units 8, 10–11
confidence limits 23
constant birth rates 113, 115–16
constants 25, 66
contingency tables 166
continuous data patterns 157–64
convection 125–6
conversion factors 8–9
correlation
  coefficients 149–50
  multivariate statistics 169, 170
  statistical tests 147, 149–50, 155–7
correspondence analysis 168
cosine (cos) functions 30–1
count data pattern 165–8
cubic scaling 42
curved functions 46, 47–53, 59–64

data
  collation 15, 20–1
  collection 15
  handling 15–19
  presentation 20–3
  sample summarizing 15–19
  transformation 143, 145
decay constants 127
decimal dilutions 14
decimal indices 35
definite integrals 88–9
degrees, angles 29–30
degrees of freedom 146–7, 159–60, 162
dependent variables 21, 46
derivatives 55–6, 58, 59–64
derived SI units 6, 7
difference equations 106–11
  algebra 28
  arithmetic sequences 106
  first-order difference equations 106, 108–11
  geometric sequences 106–8
differential calculus (differentiation) 55–64
  curved functions 59–64
  derivatives 55–6
  integration 65–6
  multiple term gradients 57–8
  second derivatives 59–64
  simple function gradients 56–7
differential equations 95–105
  algebra 28
  integration 96–9
  separating the variables 99–105
discontinuous variables 21
discrete variables 21
discriminant function analysis 169, 170
dispersion 16
division, partial fractions 81–4
double-blind methods, bias reduction 142
doubling dilutions 14

energy 9, 125–6
enzyme activity 51
equalizing variances 145
equilibrium population 117–18
error bars 23
error sum of squares 158–60
errors 4–5, 146
evolutionary phylogenies 170–1
expectation distributions 166–7
expected frequencies 165
experimental design 137, 138–42
  bias reduction 138, 140–2
  sample size 138–9
  sampling variation 138, 139–40
  statistics 138–42
exponential decay 48, 49
exponential growth 47, 48, 113, 119–20

F-ratio 144
first derivatives 60–4
first-order difference equations 106, 108–11
first-order differential equations 95
first-order polynomial functions 52
first-order reactions 130, 131–3
Fisher measure 160, 162
Fisher, Ronald 157
Fisher's exact test 165, 167
fitted line plots 21, 22
fractional indices 34–5
frequency polygon graphs 21, 22
full revolution, angles 29
functions
  curves 46, 47–53, 59–64
  straight lines 46–7

generating random numbers 142
geometric sequences 106–8
Gosset, William 157
gradients 46, 55–64
graphs
construction rules 23
data presentation 21–3
integration and integrals 87–9
log-linear graphs 33, 37–8
log-log graphs 33, 37–8
growth factors, populations 120
growth rates *see* population growth

half-life 127–8
handling data 15–19
heat loss 125–6
heterogeneity 165
hyperbolas, functions 48, 51–2
hypotenuse 30–1
hypothesis testing 143, 145–7

independent variables 23, 46, 152
indices (powers) 33–6
initial conditions, differential equations 96
integration and integrals 65–94
acceleration 71, 72–5
areas under lines 87–9
change of variables 76–8
differential equations 96–9
differentiation 65–6
methods 76–86
numerical integration 90–4
partial fractions 76, 78–86
position of a moving body 71–2
scalar multiples 67–70
standard forms 65, 66–70
sums and differences 67
velocity 71, 72–5
intercepts 46, 151
interval scales 3
inverse operations, algebra 27

kelvin units 8
kinetics 127–33
chemical reaction rates 127–33
radioactivity 127–8
Kruskal-Wallis test 164

length 9
ligand binding 52
light 8, 39, 40–1
limited growth, populations 113, 116–19, 122
line fitting 52–3
linear regression 149, 150–2
linearization, data 145
Lineweaver-Burk plots 51
location, central values 16
log-linear graphs 33, 37–8
log-log graphs 33, 37–8
logarithms 33, 35–7
logarithmic functions 48, 50
pH scales 40
logistic difference equations 117
logistic growth 48, 49
logistic model 113, 122–4
luminous intensity 8

Malthusian factor 113
manipulating numbers, algebra 25–8
Mann–Whitney *U* statistic 164
Mann–Whitney–Wilcoxon test 157, 162–4
mass 9
maxima 59
mean 16, 17, 161–2
mean squares 160, 162
measurements
accuracy 4–5
angles 29–30
bias 4
errors 4–5
methods 3
precision 4–5
SI units 6–10
types 3–5
median 16
Michaelis-Menton equation 51
minima 59
mode 16
molality 12
molar absorptivity 41
molarity 10
mole, SI units 10
molecular weight 10
multidimensional scaling 168
multiple data comparisons 157, 160–2
multiple regression 149, 152–4, 169, 170
multiple term gradients 57–8
multivariate statistics 168–71

neutrality 39
Newton's law of cooling 125–6
noise-to-signal ratio 144
nominal scales 3
nonlinear regression 149, 154–5
nonparametric tests 143, 165–7
normal distribution 143–5
normality 12, 143
normalization, data 145
null hypotheses 145–6
numerical data presentation 20–1
numerical integration 90–4

one-way analysis of variance 158–60
ordinal scales 3
ordination techniques 169

parametric tests 143
partial fractions 76, 78–86
path lengths 41
pattern searches 149–64
Pearson's correlation 147, 149
percentage solutions 13
Periodic table 11
pH scale 39–40
phylogenies 170–1
pie charts 21, 22
polynomial functions 52–3
population growth 113–24
attrition 113, 120–2
constant birth rates 113, 115–16
exponential growth 113, 119–20
limited growth 113, 116–19, 122
logistic model 113, 122–4
unlimited growth 113–14
population standard deviation 17
position of a moving body 71–2
power analysis 138–9
power functions 48, 50
powers (indices) 33–6
precision 4–5
prefixes 6, 7–8
presenting data 20–3
pressure 9
probability 135–6, 146–7
product-moment correlation tests 149
pseudoreplication 139
Pythagoras' theorem 32

quadratic functions 48, 51

radians 29–30
radioactivity 9, 127–8
random number generation 142
random variability 158–60
randomization, bias reduction 140–2
range 17
ranking 162–4
rate of chemical reactions 127, 128–33
rates of change
differential calculus 55–64
integration and integrals 65–94
ratio scales 3
rearranging algebraic equations 26–7
rectangular hyperbola functions 48, 51–2
regression 149, 150–7
relaxed definition 21
repeated factors, partial fractions 84–6
rounding errors 4

sample size 135, 136–7, 138–9
sampling errors 4
sampling variation 135, 136–7, 138, 139–40
scalar multiples, integrals 67–70
scaling 39, 41–3, 44, 169
scatterplots 21, 22
second derivatives 59–64
second-order differential equations 95, 98–9
second-order polynomial functions 52
second-order reactions 130–3

self-thinning 39, 43–5
separating the variables 99–105
sequences 106–8
serial solutions 13, 14
SI units, measurements 6–11
sigmoid functions 48, 50
significance
  levels 136, 160, 162
  linear regression 151
significant figures 5, 20
Simpson's rule 90, 91–4
sine (sin) function 30–1
sine waves 48, 50
slopes 46, 55–64, 151
software 147
solutes 10
solution preparation 10–14
Spearman's rank correlation 149
spectrophotometry 41
standard derivatives 58
standard deviations 5, 17
standard errors 17
standard integrals 65, 66–70
statistical tests 143–71
  cause and pattern searches 149–56
  chi-squared tests 165–7
  continuous data patterns 157–64
  correlation 149–50, 155–7
  count data pattern 165–8
  Fisher's exact test 165, 167
  hypothesis testing 146–7
  linear regression 149, 150–2
  Mann–Whitney–Wilcoxon test 157, 162–4
  multiple comparisons 157, 160–2
  multiple regression 149, 152–4
  multivariate statistics 168–71
  nonlinear regression 149, 154–5
  pattern searches 149–64
  presenting information 146–7
  regression 149, 150–7
  see also analysis of variance
statistics 135–71
  experimental design 138–42
  measures 15–19
  populations 16–19
  significance 136
  statistical analysis 135
  tests 143–71
stepwise regression 154
stock solutions 13–14
straight line functions 46–7
stratified sampling 139–40
strict definition, variables 21
substrate concentration 51
sum of squares 158–60, 162
summarizing sample data 15–19
summative statistics 17–19
sums and differences, integrals 67
symbols, algebra 25
Système International d'Unités (SI units) 6–11

t-test 147
tables, data representation 23
tangent (tan) function 30–1
temperature
  heat loss 125–6
  measurement units 9
  SI units 8
tests and testing 143–7
theoretical expectation distributions 166–7
third order polynomial functions 52
thresholds 136
time, units 8
total sum of squares 158–60, 162
transformation of data 143, 145
transmittance 41
trapezoidal rule 90–1, 92–4
trigonometry 29–32
  angles 29–30
  cosine function 30–1
  hypotenuse 30–1
  Pythagoras' theorem 32
  sine function 30–1
  tangent function 30–1
turning points 59
two-way analysis of variance 161–2
Type I/II errors, hypothesis testing 146

U statistic 164
units, SI units 6–11
unlimited growth, populations 113–14

variables
  algebra 25
  graphs 21–3
  integration and integrals 76–8
  separation 99–105
variance 16, 145
  see also analysis of variance
velocity 71, 72–5
volume 8, 9
volume/volume percentage solutions 13

weight 42–3, 44
weight/volume percentage solutions 13
weight/weight percentage solutions 13
Wilcoxon statistic 163–4

Yoda's law 39, 44–5

zeroth order reactions 129–30
zeroth terms 106